# THE QUANTUM WORLD

## Quantum Physics for Everyone

Kenneth W. Ford

Harvard University Press
Cambridge, Massachusetts
London, England

Drawings by Paul G. Hewitt

LIBRARY OF CONGRESS CATALOGING-IN-PUBLICATION DATA

Ford, Kenneth William, 1926–
The quantum world : quantum physics for everyone / Kenneth W. Ford.
p. cm.
Includes index.
ISBN 0-674-01342-5 (alk. paper)
1. Quantum theory.   I. Title.
QC174.12.F68 2004
530.12—dc22
2003068565

*To Charlie, Thomas, Nathaniel, Jasper, Colin, Hannah, Masha, Daniel, Casey, Toby, and Isaiah*

# Acknowledgments

Jonas Schultz and Paul Hewitt read the entire manuscript with care and made many helpful suggestions. Paul also provided the drawings that have brought ideas in the book to life. I am greatly indebted to them both. Thanks, also, to those eagle-eyed friends who read and commented on major portions (or even all) of the book: Pam Bond, Eli Burstein, Howard Glasser, Diane Goldstein, and Joe Scherrer. Diane's high-school seniors at Germantown Academy carved up the book among themselves and provided valuable (and unvarnished) feedback. They are Rachel Ahrenhold, Ryan Cassidy, Meredith Cocco, Brian Dimm, Emmanuel Girin, Alex Hamill, Mark Hightower, Mike Nieto, Luis Perez, Matt Roman, Jared Solomon, and Joseph Verdi.

Among those who provided helpful facts and data (or tried diligently to find what I was looking for) were Finn Aaserud, Stephen Brush, Brian Burke, Val Fitch, Alexei Kojevnikov, Alfred Mann, Florence Mini, Jay Pasachoff, Max Tegmark, and Virginia Trimble. Jason Ford and Nina Tannenwald helped to get Chapter 1 off on the right foot, and Lillian Lee was a valuable sounding board and source of ideas on the title. Heidi Miller Sims was my more-than-competent fact checker.

My wife, Joanne, and all of my children—Paul, Sarah, Nina, Caroline, Adam, Jason, and Ian—were unflaggingly supportive of what they long ago learned is called "work": sitting at a desk. I have had the good fortune to fall in with a most skilled and agreeable band at Harvard University Press: Michael Fisher, Sara Davis, and Maria Ascher.

# Contents

Acknowledgments     vii

1. Beneath the Surface of Things     1

2. How Small Is Small? How Fast Is Fast?     8

3. Meet the Leptons     29

4. The Rest of the Extended Family     67

5. Quantum Lumps     92

6. Quantum Jumps     112

7. Social and Antisocial Particles     131

8. Clinging to Constancy     153

9. Waves and Particles     184

10. Pushing the Limits     220

Appendix A: Measurements and Magnitudes     249

Appendix B: The Particles     252

Appendix C: Going for the Gold     257

Index     259

# chapter 1

# Beneath the
# Surface of Things

Knock on wood. It seems solid. It *is* solid. But probe more deeply and you encounter other worlds. You probably learned in school that solid matter is made of atoms and that atoms are mostly empty space. Actually, they are empty in the same sense that the disk carved out by a whirling propeller blade is empty. It's easy for something small and fast to pass through an atom or a spinning propeller, but next to impossible for something large and slow to do so.

By most measures, atoms are small.* Yet to some scientists, they are gargantuan. These scientists—nuclear and particle physicists—are concerned with what goes on in bits of space much smaller than atoms, smaller even than the tiny nuclei that sit at the centers of atoms. We call their realm of study the *subatomic world*. That's the world I want to explore in this book.

In the twentieth century we learned that nature in the subatomic

---

* How small? Ten million atoms in a row stretch less than a tenth of an inch. The scanning tunneling microscope, invented in 1981, showed for the first time the outlines of individual atoms. As late as 1900, many scientists even doubted the existence of atoms.

world behaves in weird and wonderful ways unknown in the ordinary world around us. When we look at the smallest specks of space and the tiniest ticks of time, we see what can only be called fireworks. Myriad new particles pop into existence, some long-lived, most short-lived, each of them interacting in some way with every other, and each capable of being destroyed as well as created. In this world, we come face to face with a speed limit in nature, find space and time stirred together, and learn that mass can change to energy and energy to mass. The strange rules of the game in this world stretch the minds of scientists and nonscientists alike.

These rules are the product of two great revolutions in twentieth-century physics, called *quantum mechanics* (roughly, the physics of the very small) and *special relativity* (roughly, the physics of the very fast).

In this book I want to explain how these two revolutionary theories—and especially quantum mechanics—have changed the way we view the world. To illustrate the ideas, I'll enlist the help of the subatomic particles (henceforth I'll just call them *particles*), the "things" that are governed by the quantum rules. After a look at "how small is small" and "how fast is fast," I'll introduce you to quite a few members of the particle families. Then I'll turn to the marvelous array of ideas that physicists have developed to account for the behavior of the particles and all that is built of particles.

In 1926, the year I was born, the only known inhabitants of the subatomic world were the electron and the proton. The *electron* is a speck of negative electricity that zips around inside atoms and migrates along wires carrying electric current. Nowadays, it is also steered magnetically toward the pixels on the screens of cathode-ray tubes that serve as computer monitors and television tubes, causing them to shine in words and pictures. The *proton*, nearly 2,000 times more massive than the electron and carrying a positive charge, sits alone at the center of the lightest hydrogen atom, pulling on the electron that circles it. Protons were, in the 1920s, assumed to exist in heavier atomic nuclei as well, and we know now that indeed they do. They also come barreling toward Earth in great numbers and with great energy from outer space, constituting what we call *primary cosmic radiation*.

The *photon*—the particle of light—was also known in 1926, but it wasn't considered a "real" particle. It had no mass; it couldn't be slowed down or corralled; and it was all too easily created and annihilated (emitted and absorbed). It wasn't a reliable, stable chunk of matter like the electron or the proton. So, even though the photon surely behaved like a particle in some respects, physicists fudged by calling it a "corpuscle" of light. Only a few years later it gained full and equal status as a real particle, when physicists realized that electrons could be created and annihilated every bit as easily as photons, that the wave properties of electrons and the wave properties of photons were all pretty much of a piece, and that a particle with no mass was, after all, a perfectly ordinary particle.

The year 1926 was right in the middle of a golden age of physics. In a brief few years, 1924 to 1928, physicists came up with some of the most important—and startling—ideas that science has ever known. These included the discovery that matter, not just light, has wave properties; the realization that the fundamental laws of nature are laws of probability, not laws of certainty; the understanding that there is a limit in principle to the accuracy with which certain properties of matter can be measured; the discovery that electrons spin about an axis that can point in only two possible directions, "up" and "down"; the prediction that for every particle there is a companion antiparticle; the insight that a single electron or photon can be moving in two or more different ways at the same time (as if you could be driving due north and due west simultaneously, or window shopping in New York and Boston simultaneously); and the principle that no two electrons can be in the same state of motion at the same time (they are rather like marchers who, try as they might, can't march in step).

These are among the "big ideas" that are central to this book. Using the subatomic particles as illustrations will help to bring these ideas to life. Particles (and, to some extent, whole atoms) are the entities most conspicuously influenced by the laws governing the very tiny and the very speedy.

I should note right here that in quantum physics, separating what *is* (the particles) from what *happens* (the laws) is not so easy. In the "classical" physics developed in the three centuries before the twentieth, the separation between what is and what happens is pretty clean. The Earth (what is) follows an orbital path (what happens) about the Sun in response to laws of force and motion. What the Earth is made of, whether it contains life or doesn't contain life, whether it spews out lava or lies dormant—these are features that have nothing to do with how the Earth moves around the Sun. To take another example: An oscillating electric charge generates electromagnetic radiation. The radiation "cares" not in the least whether the charge is carried by electrons, protons, ionized atoms, or tennis balls. It "knows" that some electrified thing is oscillating in a certain way, but doesn't "know" or need to know what that thing might be. The nature of the oscillating object (what is) has nothing to do with the radiation that is emitted (what happens).

For the particles, things are not so simple. What the particles *are* and what they *do* are intertwined. That's all part of the strangeness of the subatomic world. So, in the following chapters, you (and I) will have to keep an eye out for times when the *properties* of the particles get mixed up with the *actions* of the particles.

Let me also pause to ask *why* the subatomic world is so strange, *why* it is so weird and wonderful. Why do the laws governing the very small and the very swift defy common sense? Why do they stretch our minds to the limit? Their strangeness could not have been predicted. The classical scientists (pre-1900) assumed, rather naturally, that ordinary conceptions from the world around us, the world of our senses, would continue to serve us as knowledge accumulated about realms of nature beyond the range of our senses—about things too small to touch and too fast to glimpse. On the other hand, those classical scientists had no way of knowing that the rules would stay the same. How could they be sure—how can *anyone* be sure—that the "common sense" derived from ordinary observations will serve to account for phenomena that can't be seen, heard, or touched?

In fact, the physics of the past hundred years has taught us that common sense is a poor guide in the new realms of knowledge. No one could have predicted this outcome, but no one should be surprised by it. Everyday experience shapes your opinions about matter and motion and space and time. Common sense says that solid matter is solid, that all accurate watches keep the same time, that the mass of material after a collision is the same as it was before, and that nature is predictable: sufficiently accurate input information yields reliable prediction of outcomes. But when science moves outside the range of ordinary experience—into the subatomic world, for instance—things prove to be very different. Solid matter is mostly empty space; time is relative; mass is gained or lost in a collision; and no matter how complete the input information, the outcome is uncertain.

Why is this? We don't know why. Common sense could have extended beyond the range of our senses, but it didn't. Our everyday worldview, it turns out, is a limited one, based on what we directly perceive. We can only echo the parting words of the respected old TV news anchor Walter Cronkite: "That's the way it is." You can be enchanted, you can be amazed, you can be befuddled, but you shouldn't be surprised.

By the time I was fifty, in 1976, the known subatomic particles numbered in the hundreds. A few were added in the 1930s, a few more in the 1940s, then a flood in the 1950s and 1960s. Physicists had stopped calling the particles "elementary" or "fundamental." There were just too many of them for that. Yet just as the number of particles seemed to be getting out of hand, physicists were coming up with a simplifying scheme. A manageably small number of particles appeared to be truly fundamental (including quarks, which, to this day, no one has seen directly). Most of the known particles, including the old familiar proton, were composite—that is, built from combinations of the fundamental particles.

We can see an analogy here to what had happened decades earlier with our understanding of atoms and nuclei. By the time the *neutron* (an uncharged, or neutral, sibling of the proton) was discovered in 1932, the number of known atomic nuclei had grown to several hundred. Each was characterized by a mass and a positive charge. The charge determined the *atomic number*, or place in the periodic table. In other words, it defined the *element* (an element is a substance with unique chemical properties). The hydrogen nucleus had a charge of one unit, the helium nucleus a charge of two units, the oxygen nucleus a charge of eight units, the uranium nucleus a charge of ninety-two units, and so on. Some nuclei with the same charge (therefore nuclei of the same element) had different masses. Atoms built around these nuclei were called *isotopes*. Scientists felt sure that these several hundred nuclear types, of ninety or so elements averaging two or three isotopes each, were built of a smaller number of more fundamental constituents; but prior to the discovery of the neutron, they couldn't be sure exactly what those constituents were. The neutron made it all clear (though later it, too, was discovered to be composite). Nuclei were constructed of just two particles, the proton and the neutron. The protons provided the charge, and the protons and neutrons together provided the mass. Whizzing around the nucleus in the much larger volume of the whole atom were electrons. So just three basic particles accounted for the structure of hundreds of distinct atoms.

For the subatomic particles, the "discovery" of quarks played a role similar to the discovery of the neutron for atoms. I put "discovery" in quotation marks because what Murray Gell-Mann and George Zweig, both at Caltech, did independently in 1964 was to *postulate* the existence of quarks, not *prove* their existence through observation (the name "quark" we owe to Gell-Mann). The evidence for quarks, though still indirect, is by now overwhelming. Today, quarks are recognized as the constituents of protons, neutrons, and a whole host of other particles.

Physicists then invented the *standard model* of the subatomic particles. In this model there are twenty-four fundamental particles, including the electron, the photon, and half a dozen quarks, accounting for all observed particles and their interactions.* Twenty-four is not as pleasingly small a number as three (the number of particles known in 1926), but so far these twenty-four remain stubbornly "fundamental." No one has found any of them to be made of other, more fundamental entities.

---

\* These twenty-four do not include the graviton—the hypothetical particle of gravity—or another hypothetical entity, the Higgs particle (the only member of the particle zoo named after a person). Also excluded from the count are the antiparticles.

Murray Gell-Mann (b. 1929), 1959. Photo by Harvey of Pasadena; courtesy of AIP Emilio Segrè Visual Archives.

But if the superstring theorists have it right (I'll discuss their ideas later), there may be smaller, simpler structures that await discovery.

Some of the fundamental particles are called *leptons*, some are called *quarks*, and some are called *force carriers*. Before I introduce you to them, let's have a look at quantities and magnitudes typically used to describe what goes on in the subatomic world.

## chapter 2

# How Small Is Small?
# How Fast Is Fast?

How big—or how small—are the things we measure in the subatomic arena? You know that atoms are tiny and subatomic things even tinier, that light moves at enormous speed, that particles fly nearly as fast, and that the blink of an eye is vastly longer than the typical lifetime of a particle. It's easier to say these things than to visualize them. My purpose in this chapter is to help you "see" the subatomic realm, so that you begin to feel comfortable with the small sizes, the high speeds, and the brief time spans.

It turns out that most of the concepts needed to describe particles aren't strange at all; they are merely different in scale. Length, speed, time, mass, energy, charge, and spin can be used to describe a bowling ball as well as an electron. In the subatomic realm, the questions are: How big are these quantities? How do we know? What are convenient units in which to measure them?

To deal with the large and the small, we need a simple noncumbersome notation. Many readers may know the notation already. One thousand is 1,000, or $10^3$. One million is 1,000,000, or $10^6$. One billion is 1,000,000,000, or $10^9$. It looks simple. The power of ten is the number of zeros when the number is written out. But it's better to think of the power of ten as shorthand for the number of places the decimal point is moved. Thus 243 million, or 243,000,000, become $2.43 \times 10^8$. From 2.43 to 243,000,000, the decimal point is moved 8 places to the right. The shorthand involving powers of ten is called *exponential notation* (or, often, *scientific notation*).

The rules for numbers smaller than 1 are similar (in fact, they are really the same). One thousandth is 0.001, or $10^{-3}$ (if you start with 1 and move the decimal point three places to the left, you get 0.001). Suppose a large molecule has a length of 2.2 billionths of a meter. That means its length can be written 0.0000000022 m, or, much more conveniently, $2.2 \times 10^{-9}$ m.

*Multiplication* in scientific notation is carried out by *adding* exponents. One billion is one thousand times one million: $10^3 \times 10^6 = 10^9$. One trillion is one thousand times one billion: $10^3 \times 10^9 = 10^{12}$. To *divide*, you *subtract* exponents. For instance, what is the speed of a particle that covers the $8 \times 10^{16}$ meters to a not-so-distant star in $4 \times 10^8$ seconds (about thirteen years)? It is the distance divided by the time: $8 \times 10^{16}$ meters divided by $4 \times 10^8$ seconds $= 2 \times 10^8$ m/s (about 2/3 the speed of light).

Scientific notation is one way that scientists cope with large and small numbers. The other way is to introduce new units more appropriate to the domain being studied. We do this in the large-scale world, where (at least in the United States and Great Britain) we are likely to measure our height in feet and inches and our travel distances in miles. Astronomers pick a much larger unit, the light-year,* for measuring the distance to other stars. In our computers, we use the units kilobyte, megabyte, and gigabyte ($10^3$, $10^6$, and $10^9$ bytes, respectively). Pharmacists use milligrams (thousandths of a gram). Jet pilots use Mach number (speed expressed as a multiple of the speed of sound in air).

In the world of particles, we use the femtometer† ($10^{-15}$ m) as a convenient unit of length, the speed of light ($3 \times 10^8$ m/s, designated $c$) as a

---

* Or sometimes the parsec, which is 3.26 light-years.

† The femtometer used to be called a "fermi," in honor of the great Italian-American nuclear physicist Enrico Fermi. What luck that the abbreviation for "femtometer" is fm.

convenient unit of speed, the magnitude of the electron's charge (called *e*) as a convenient unit of charge, and the electron volt (eV) as a convenient unit of energy. The electron volt is the energy acquired by an electron being accelerated through an electric potential (or voltage) of 1 volt. An electron in a television tube that gets an electric push (or pull) of 1,500 volts, for example, flies toward the screen with an energy of 1,500 eV. One electron volt is also roughly the energy carried by a photon of red light. Ten electron volts is about what it takes to *ionize* an atom (kick an electron free). Because of the equivalence of mass and energy, particle masses can also be expressed in eV units. The electron, for instance, has a mass of 511,000 eV ($511 \times 10^3$ eV, or 511 keV) and the proton has a mass of 938,000,000 eV ($938 \times 10^6$ eV, or 938 MeV, just shy of 1 GeV).* Note that particle masses are relatively large in the eV unit.

Table A.2 (in Appendix A) shows some properties of matter with common units that are used in the large-scale world and typical magnitudes in the subatomic world.

## Length

Is length a dull subject? Not when you start trying to picture subatomic distances. Atomic nuclei range from about $10^{-4}$ to $10^{-5}$ of the size of an atom. Imagine an atom expanded to a diameter of 3 km, about as big as a medium-sized airport. A fraction $10^{-4}$ of this is 30 cm, or about one foot, roughly the diameter of a basketball. A basketball in the middle of an airport is as lonely as a "large" nucleus such as that of uranium at the middle of its atom. (To make the model more apt, cover the airport with a giant dome that reaches, at its center, a mile above the ground.) Now replace the basketball with a golf ball and you have a model of the hydrogen atom, with its central proton.† Inside the golf ball (the proton),

---

* Table A.1 in Appendix A gives the standard names and symbols for large and small multipliers.

† It turns out that a uranium atom is scarcely any larger than a hydrogen atom (so no need to change the size of the airport). The more highly charged uranium nucleus is pulling harder on its electrons, but it has more electrons to pull, a balancing act that results in little change of the atom's size.

quarks and gluons are buzzing about, none of which have any discernible size. And outside the golf ball, a single electron, similarly lacking discernible size, "fills up" the 3-km sphere of the atom. An invisible speck filling a huge volume? Yes. Thanks to the wave nature of matter, it is so.

With this magnification that fattens a proton to the size of a golf ball, what happens to a child's marble, with a diameter of 1 cm? It expands to a sphere with a diameter equal to the diameter of Earth's orbit. Inside the expanded marble are the Sun, Mercury, and Venus, while Earth traces a circle on the marble's surface.

Our golf ball, shrunk back to the actual size of the proton, has a diameter of about $10^{-15}$ m. This is equal to 1 femtometer, or 1 fermi (1 fm). The smallest distance probed in any experiment so far conducted is about one thousandth of a fermi, or $10^{-18}$ m. The fundamental particles, if they have any size at all, are smaller than this.

How do we measure such incredibly small distances? Not with a ruler, not with a pair of calipers. One way is through *scattering experiments*. Let's say that a beam of electrons is fired at a sample of hydrogen, which contains lots of protons. The electrons are deflected—that is, scattered—by the electric force acting between the electron projectiles and the proton targets. If the protons were point particles (so small that they could be considered to exist at mathematical points), the electrons would emerge with a certain predictable pattern. The way in which the *actual* pattern of scattered electrons differs from such a prediction reveals at what distance of approach to a proton an electron begins to feel a force different from that expected for point particles. Analysis of the pattern reveals the size of the proton to be about 1 fm, and even reveals how the positive electric charge of the proton is distributed within the proton. (It was through scattering experiments with alpha particles that Ernest Rutherford first revealed, in the 1920s, that atomic nuclei have finite size.)

The wave nature of matter also helps in the measurement of tiny distances. Every moving particle has a certain wavelength. The more energetic the particle, the shorter its wavelength. Waves *diffract* from targets and can reveal features of the targets that are as large as or larger than the wavelength. Water waves, for instance, can reveal features of an anchored ship past which they roll, but tell little or nothing about

Figure 1. The passing waves reveal much about the ship, but nothing about the chain.

an anchor chain they pass. Light waves, with the help of the finest microscopes, can reveal features of biological samples as small as the wavelength of the light, but not features such as viral structure that are smaller than the light's wavelength. The most energetic particles emerging from modern accelerators have wavelengths around 0.001 fm ($10^{-18}$ m). Their diffraction probes down to distances that small.

The universe at large is not the subject of this book, but it's of some interest to see where we humans fit between the largest and smallest regions that scientists have explored. The radius of the known universe is about fourteen billion light-years, which works out to about $10^{26}$ m. The width of a desk or table is about $10^{26}$ times smaller than the universe and "only" $10^{18}$ times larger than the smallest distances probed in particle experiments. A "mean" distance between the large and small limits is $10^{4}$ m, or 10 km (some six miles). Let's say that's the distance you commute each day. Multiply that distance by $10^{22}$ and you have the radius of the universe. Divide it by $10^{22}$ and you have the current small-scale limit of exploration. It may be of some satisfaction to know that a nice familiar distance like six miles lies midway (expressed as a ratio) between the largest and smallest distances currently known to science. How big is that number $10^{22}$? Well, if you commuted for that many days, your working life would be about two billion times longer than the age of the universe. If you spread that many people around the galaxy, it would take more than a trillion planets to hold them. It's also, in round numbers, the total number of stars in the known universe and the number of atoms in a breathful of air. How about $10^{44}$, the ratio of the largest to the smallest distances? I leave it to the reader to think of some way to bring home the enormity of that number.

Before leaving the topic of length, I have to mention an almost unimaginably small distance studied by theorists. It is the so-called *Planck length*,* about $10^{-35}$ m. Recall that a proton's size is about $10^{-15}$ m. This makes the Planck length some $10^{20}$ times smaller—a hundred billion billion times smaller—than the already minuscule proton. One must marvel that physicists dare to ponder such a domain, in which, according to calculations, not just particles but space and time themselves become subject to the strange rules of quantum mechanics. At this Planck scale of size, space and time are expected to lose the smooth predictable character they have in our everyday world and degenerate into a roiling *quantum foam*. And at this same scale of size, hypothetical *strings* may do their hula-hoop dances, giving rise to the particles we see in the laboratory.

## Speed

A snail in a hurry can travel at a speed of about 0.01 meter per second (0.01 m/s). A person strolls at about 1 m/s, drives a car at 30 m/s, and rides in an airplane at close to the speed of sound, which (in cool air) is 330 m/s (740 miles per hour). Light makes its way from place to place almost a million times faster, at $3 \times 10^8$ m/s.

We know of no firm limit, large or small, in distance. But nature has established a firm speed limit, the speed of light. No speeding ticket has ever been issued for exceeding the speed of light, for, so far as we know,

---

\* The American physicist John Wheeler, noted for his work in nuclear physics and gravitation, coined the term *Planck length* in the 1950s as a tribute to the German quantum pioneer Max Planck. (Many of Wheeler's coinages have caught on in physics. Among them are *quantum foam* and *black hole*.)

nothing has ever managed to crack that barrier. Even an astronaut in orbit falls short of the speed of light by a factor of forty thousand. The astronaut needs an hour and a half to get once around the Earth, whereas light, if steered around in an optical fiber, could do the circumnavigation in just over a tenth of a second. Still, we are not so far removed from nature's top speed as we are from the frontiers of time and distance.

In 1969, people for the first time had the opportunity to sense the finite speed of light (or, what is the same thing, the speed of radio waves). We heard a NASA controller in Houston speak to an astronaut on the moon. Then, with a delay noticeably longer than would be expected in an ordinary conversation, we heard the astronaut's response. The time delay, in addition to normal reaction time, was the time it took the signals to get to the moon and back at the speed of light, about 2.5 seconds. Children learn in school that light from the Sun takes eight minutes to reach Earth. Starlight from the next-nearest star takes four years to reach us, and light from the remotest reaches of the universe takes more than ten billion years to get here. To the astronomer, light just ambles along.

Atoms and molecules, in their restless motion in solids, liquids, and gases on Earth, get around at one to ten times the speed of sound in air, a factor of $10^5$ to $10^6$ short of the speed of light. Yet for particles in accelerators or arriving from outer space as cosmic rays, speeds close to the speed of light are commonplace. Photons, being massless, have no choice. They travel at the only speed available to massless particles, the speed of light. They *are* light. Neutrinos, with their wispy mass, move almost as fast as photons. In modern accelerators, electrons, too, zoom at phenomenal speed, being slower than light by the speed of a strolling bug.

Measuring speed, even for the fastest particles and for light itself, is straightforward: distance divided by time, just as for everyday motion. Modern clocks can slice time to billionths of a second or less. In one billionth of a second, or 1 nanosecond, light travels about one foot (30 cm). The speed of light is, in fact, so well measured that it has been adopted as a fixed standard, making time and space measurements secondary.

When in a hurry, the crew of the Starship Enterprise shift into warp drive and scoot about the galaxy at speeds well beyond the speed of light. Is there any chance that such science-fiction speeds will become reality? It is extremely unlikely, and for a simple reason. The lighter an object is, the more easily it can be accelerated. Freight trains lumber slowly up to speed, automobiles more quickly, and protons in an accelerator still more quickly. A particle with no mass at all is the easiest to accelerate; indeed, the massless photon jumps instantaneously to the speed of light when it is created. But not beyond. If anything at all were able to go faster than light, then light itself, being composed of massless photons, should go faster. But scientists resist absolutes. A hypothetical entity called a *tachyon*, which *can* go faster than light, has been studied theoretically. It's a weird entity, for in some frames of reference it can arrive at its destination before it starts its trip. Nevertheless, theorists grit their teeth and study it. Searches for the tachyon have, to date, found nothing.

## Time

What is a "short" time and what is a "long" time? For us humans, a year is a long time and a hundredth of a second is a short time.* For a particle, on the other hand, a hundredth of a second is an eternity. For the stately march of events in the cosmos, a million years is something like a lunch break.

A good choice for the "tick" of a particle's clock is the time it takes the particle, moving at close to the speed of light, to cross the diameter of a proton. That's about $10^{-23}$ s, much less than a billionth of a billionth of a second. A gluon (the "glue" particle within a nucleus) lasts about that long between its creation and annihilation. A pion (a particle created in nuclear collisions) that moves a whole foot has traveled nearly a million billion times the diameter of a proton and has taken a lengthy $10^{-9}$ second to do it. Particles that live long enough to leave tracks in a detector have lifetimes of $10^{-10}$ to $10^{-6}$ s. The neutron is a

---

* You can perceive an image flashed for a hundredth of a second but not a thousandth. Some Olympic events are decided by a few hundredths of a second.

strange special case. With an average lifetime of fifteen minutes, it is the Methuselah of the particle world.

Since the shortest distance ever probed experimentally is about $10^{-18}$ m, it is fair to say that the shortest time studied is about $10^{-26}$ s (although direct measurements of time are still very far from reaching this short an interval).* The longest known time is the "lifetime of the universe"—that is, the apparent duration of the expansion of the universe, currently estimated to be 13.7 billion years, or nearly $10^{18}$ s. The ratio of these times is $10^{44}$, the same enormous number as the ratio of the largest to the smallest known distances. This is not a coincidence. The outermost regions of the universe are moving away from us at a speed near the speed of light, and the particles flying about in the subatomic world are also moving at such a speed. On both the cosmological and subatomic frontiers, the speed of light is the natural link between distance and time measurements.

## Mass

Mass is a measure of *inertia*—of how hard it is to set a stationary object into motion or to stop or deflect a moving object. You can stop a thrown baseball without too much trouble. To stop a bowling ball coming at you with the same speed would be much harder. Don't even think about trying to stop a freight car bearing down on you with the same speed. The bowling ball has more mass, or inertia, than the baseball, and the freight car has still more. The greater the mass, the harder it is to change the object's *state of motion*. Subatomic particles, relative to ordinary objects, have almost no mass at all. Every minute, you are stopping lots of muons that barrel into your body, and you don't feel a thing. (*Muons* are unstable particles created high in the atmosphere by incoming cosmic rays. Streaking downward, they form part of what is called *background radiation*.)

In the everyday world, we think of mass as heaviness. We find the mass of an object by weighing it. By chance (not really by chance—

---

* Cosmologists have dared to push their calculations back to times as early as $10^{-43}$ seconds after the Big Bang.

there's a deep reason for it), gravity pulls on an object with a force proportional to its mass. So, at the surface of the Earth, we can determine the mass of an object by measuring how strongly it is pulled downward by Earth's gravity. This works fine in grocery stores and at truck weigh-stations, but it doesn't work in outer space. An astronaut on board a circling space station is weightless, but still has mass. If astronaut Julie or astronaut Jack stands on a scale while they are orbiting, the scale reads zero. That's what we mean by weightless. But if Julie and Jack join hands and push apart, each must exert some effort to cause motion. That's because each has mass, or inertia. The speeds at which they drift apart are in inverse proportion to their masses. If, after they push each other, Julie drifts at a speed of 1.2 m/s and Jack drifts at a speed of 1.0 m/s, it's because Jack's mass is 1.2 times greater than Julie's. He resists being set into motion by that much more. Julie could measure her own mass by tossing a 1-kg weight and finding out how much faster the weight flies away than she recoils. In practice, to find out whether an astronaut on a long mission is gaining or losing weight, the astronaut is shaken side to side in a specially designed chair. A mechanism in the chair measures how much resistance the astronaut offers to being shaken, and translates the measurement into a "weight" (really a mass).

The tiny mass of a particle can be measured in a similar way. If the particle has charge, it can be deflected by a magnetic field. Given knowledge of the particle's speed, the scientist can infer the particle's mass from the curvature of its track. We even speak of the *rigidity* of a particle in motion, meaning its resistance to having its direction of motion changed. A particle with either more mass or more speed has greater rigidity.

Einstein's mass-energy equivalence ($E = mc^2$) also provides a way to measure particle masses. I will discuss this equation and some of its implications in the next section. For now, suffice it to say that an experimenter who knows the total energy of a particle—for instance, by knowing how it was created—and measures its kinetic energy (energy of motion) can subtract the kinetic energy from the total energy to get the mass energy, and thereby the mass.

As pointed out in Chapter 1, energy is commonly used to gauge and report the masses of particles. When we say, for instance, that a proton's mass is 938 million electron volts (938 MeV), we mean that the proton's mass, multiplied by the square of the speed of light, is this much energy. In kilograms, the proton's mass is a minuscule $1.67 \times 10^{-27}$ kg. You can see why it is handier to use MeV or GeV than kg to report particle masses.

To end the discussion of mass on a cosmic note, let's ask: What is the mass of the universe? This is certainly not well known, but a rough estimate can be made. Astronomers believe that there are about $10^{22}$ stars in the visible universe (slightly fewer than the number of molecules in a gram of water). An average star weighs (that is, has a mass of) about $10^{30}$ kg, making the mass of stars some $10^{52}$ kg. Each kilogram of matter contains about $10^{27}$ protons, so the visible universe contains (very roughly) $10^{79}$ protons.* There is also an invisible universe (the so-called *dark matter*), which may be six times more massive than the visible universe. What is dark matter? That is one of the great unanswered questions of present-day cosmology.

## Energy

Like an actor who disappears from the stage wearing one costume and reappears wearing another, energy has many guises and can swiftly change from one to another. Thanks to its rich diversity of form, energy appears in nearly every part of the description of nature and can make a good claim to be the most important single concept in science.

The importance of energy springs not just from its variety of form, but from its conservation: the total amount of energy in the cosmos remains always the same, since the loss of one kind of energy is always being compensated by the gain of another kind of energy. We speak of potential energy, chemical energy, nuclear energy, electrical energy, radiant energy, heat energy, and more. In the particle world, there are just two significant forms of energy: kinetic energy and mass energy. *Kinetic energy* is energy of motion; *mass energy* is energy of being.

---

* The known part of the universe presumably contains a like number of electrons, about $10^{79}$. It contains about a billion times more photons, roughly $10^{88}$, and nearly as many neutrinos as photons. (The physicist George Gamow used to carry around a small matchbox labeled, "Guaranteed to contain at least 100 neutrinos.")

The faster a particle moves, the more kinetic energy it has. At rest, its kinetic energy is zero. Of two particles moving with the same speed, the one of greater mass has greater kinetic energy.* Photons are oddities. They move at an invariable speed $c$, and possess kinetic energy despite their zero mass. Since they can't be slowed or stopped (although they can be destroyed), they can never have zero kinetic energy. And since they have no mass, they have no mass energy. They are purely kinetic creatures.

It was Albert Einstein, working early in the twentieth century, who discovered that mass is a form of energy. Everyone knows the formula

$$E = mc^2.$$

Let's see just what it means. First, it tells us that mass energy, or energy of being, is proportional to mass. Twice as much mass means twice as much mass energy, and no mass means no mass energy. The quantity $c^2$, the square of the speed of light, is called a *constant of proportionality*. It does the job of converting from the unit in which mass is expressed to the unit in which energy is expressed. By analogy, consider the equation giving the cost of filling your car with gasoline,

$$C = GP.$$

The cost, $C$, is equal to the number of gallons, $G$, multiplied by the price per gallon, $P$. The cost is proportional to the number of gallons and $P$ is a constant of proportionality, which converts the number of gallons to the number of dollars. In a similar way, $c^2$ is a price. It is energy per unit mass, the price that must be paid in energy to create a unit of mass. And a high price it is. In standard units (energy in joules, mass in kilograms): $c^2 = 9 \times 10^{16}$ J/kg. So mass represents a highly concentrated form of energy. A little mass yields a lot of energy. A lot of energy is needed to make a little mass.

---

* For speeds much less than the speed of light, the kinetic energy of a particle of mass $m$ moving with speed $v$ is expressed by KE $= (1/2)mv^2$. At speeds near the speed of light, this "classical" formula is replaced by a "relativistic" formula according to which KE approaches an infinite value as the speed approaches the speed of light. Light itself obeys yet a different formula. Its KE depends on its frequency or wavelength. A blue photon has more KE than a red photon.

Albert Einstein
(1879–1955),
1954. Photo by
Richard Arens;
courtesy of AIP
Emilio Segrè
Visual Archives.

A major purpose of modern accelerators is to change kinetic energy into mass energy. When a proton, with a kinetic energy that is perhaps a thousand times its rest energy, slams into another proton, a great deal of that kinetic energy is available to make new mass. Dozens or hundreds of particles may fly away from the point of collision. Analysis of what happens in such a collision is made possible by applying the laws of conservation of energy and momentum.*

Given energy's manifold forms, it's not surprising that a great many different units of measurement for it have sprung up. One joule is the kinetic energy of a 2-kg mass moving at 1 m/s. One calorie is the energy needed to raise the temperature of one gram of water by one Celsius degree. It's equal to about four joules. One food calorie (or "large calorie," or just Calorie with a capital C) is one thousand calories. Two thousand to three thousand food calories are needed to keep the human machine running for a day. Another unit, the kilowatt-hour, is what appears on monthly electric energy bills. One kilowatt-hour, equal to 3.6 million joules, will light a 100-watt bulb for ten hours.†

An agreeable feature in the subatomic world is that energy and mass are typically measured in the same unit, the electron volt. The earliest cyclotrons, in the 1930s, accelerated particles to energies of a few million electron volts (MeV). Over the decades that followed, accelerator energies climbed upward, to hundreds of MeV, then to many GeV, and now to more than a TeV (tera electron volt, or trillion electron volts). Accelerator energies vastly exceed the thermal energies of even the hottest materials. At the surface of the Sun, protons move with kinetic energies of only about 1 eV, a billion times less than the energy locked within the mass of the proton. But in a TeV accelerator, the proton gains a kinetic energy a thousand times its rest energy. Some cosmic rays bombarding Earth from outer space have energies up to $10^{20}$ eV, enor-

---

* *Momentum* is a directed quantity (or vector quantity), meaning that it points in the direction of motion of a particle. In classical physics, momentum is $mv$, the product of mass and velocity. Isaac Newton called it the "quantity of motion." For high-speed particles, momentum is $mv / \sqrt{1 - (v/c)^2}$, where $c$ is the speed of light. A momentum pointing north and an equal momentum pointing south "add" to zero.

† Newspapers all too often confuse energy and power. Power, measured in watts, is energy per unit time. Conversely energy—what your power company sells to you— is power multiplied by time (as in kilowatt-hours).

mously greater than the highest particle energies achieved on Earth. Where these particles get their energy remains a mystery.

## Charge

Electric charge is that certain something (that *je ne sais quoi*) which makes a particle attractive to the opposite kind of particle. Neutral particles, lacking charge, fail to attract other particles (at least electrically). Charge can lead to pairing. The hydrogen atom, for example, consists of an electron and a proton held together by electrical attraction. More energetic particles don't pair up as the result of electric forces; they merely deviate from a straight path.

Like charges (both positive or both negative) repel; unlike charges (positive and negative) attract. Within a nucleus, the positively charged protons repel one another, but the attractive "glue" provided by the gluons overcomes this repulsion and holds the nucleus together—up to a point. Eventually, for very heavy nuclei, the electrical force becomes more than the gluons can counteract, and the nucleus flies apart. That's why no nuclei heavier than uranium nuclei exist in nature.

For the large-scale chunks of matter in the universe (planets, stars, and galaxies), gravity—intrinsically weak but always attractive—prevails. On a smaller scale, it's easy to see electricity outpulling gravity. Run a comb through your hair and then use it to lift a bit of paper from a table. A charge imbalance in the comb of perhaps one part in a million million ($1$ in $10^{12}$) is enough to make the comb pull up electrically on the bit of paper more strongly than Earth's gravity pulls down on it. If the comb were somehow to acquire a more substantial imbalance of charge, either a deadly bolt of lightning would jump from the comb to your head as the charge neutralized, or the comb would become a lethal weapon as it was tugged mightily toward your head.

Which charge is called positive and which negative is entirely arbitrary, and is the result of a historical accident. In the mid-eighteenth century, Benjamin Franklin reached the conclusion that one kind of electricity flows readily from one object to another; this kind came to be called positive. And so, eventually, protons came to be called positive and electrons negative. Now we know that it is the negative electrons that are mobile and account for the flow of electric charge in metals.

Charge is almost as mysterious as the first paragraph of this section suggests. Physicists understand its *conservation* (related to a subtle mathematical symmetry in nature), but they don't really understand its *quantization*. Why are the magnitudes of the proton and electron charge exactly equal, as they appear to be? Why do all the particles we observe have a charge equal to that of the proton ($+1$) or its negative ($-1$) or some simple multiple thereof (such as $+2$ or $-2$), whereas the quarks have fractional charges (such as $+\frac{2}{3}$ and $-\frac{1}{3}$)? What happens in the immediate vicinity of a charged fundamental particle? If the particle is a true point, with no physical extension, the electric field of force becomes infinitely large at the particle.* If the particle has physical extension, why don't the bits of charge that make up its total charge cause it to fly apart? Are space and time drastically warped right at the particle's location? These may not even be the right questions to be asking, but they suggest why we can't say that we fully and deeply understand charge.

The electron deserves a monument in its honor, for it is not only a very special fundamental particle (the lightest one with charge)—it is also the workhorse of both heavy industry and the communications industry. Electrons are pushed and pulled through the tiny circuits in computers; they are set oscillating in the antennas that send and receive wireless messages; when pulsing through wires they can turn powerful motors; and when they encounter impediments to their motion, they emit light and heat.

In our everyday world, the unit of charge is the coulomb (named for the French scientist Charles A. Coulomb, who in 1785 discovered the exact law of electric force). One coulomb is roughly the amount of charge that moves through a 100-watt light bulb in a second. The charge unit in the subatomic world is, of course, much smaller. As given in Table A.2, this unit $e$ is equal to $1.6 \times 10^{-19}$ coulomb (less than a bil-

---

* According to the inverse-square law, the field of force gets larger and larger as the distance from the particle gets smaller and smaller, reaching infinity when the distance reaches zero.

lionth of a billionth of a coulomb). Turning the numbers around, we can say that some $6 \times 10^{18}$ electrons flow through the light bulb each second.

## Spin

Just about everything spins, from photons and neutrinos to galaxies and clusters of galaxies. Our Earth rotates on its axis once a day and revolves once around the Sun in a year. The Sun itself turns on its axis once every twenty-six days and circles the galaxy in 230 million years. On an even longer time scale, galaxies revolve about one another in clusters. The question "Does the universe as whole rotate?" is not really meaningful, for one would have to ask, "Relative to what?" The famous logician Kurt Gödel once got interested in a related question: Do more galaxies rotate in one direction than another? As far as he could tell from available data, the axes of galactic rotation are randomly distributed in all directions. In that sense, the universe as a whole apparently does not rotate—at least, not enough for us to detect.

Going down the scale of size, molecules rotate (at rates dependent on temperature), and electrons within atoms circle the atomic nucleus at speeds of from 1 percent to more than 10 percent the speed of light. Nuclei can rotate—most do—and the protons, neutrons, quarks, and gluons within the nucleus all spin as well. Indeed, most particles, fundamental as well as composite, have this property of spin.*

At every scale of size, it is useful to distinguish two kinds of rotational motion. One is the turning of an object about its own axis (such as the daily rotation of the Earth), which we call *spin*. The other is the circling of an object about some other point (such as the yearly revolution of the Earth around the Sun), which we call *orbital motion*. Both kinds of rotation are measured in terms of *angular momentum*, a combined measure of the mass, size, and speed of the rotating system. Angular momentum measures the "strength" or "intensity" of rotational mo-

---

* When the subject of spin was driving my students in a high-school physics course to distraction, I rallied their spirits by distributing letterhead emblazoned with the motto Παντα κυκλει (*Panta kuklei*, "Everything rotates").

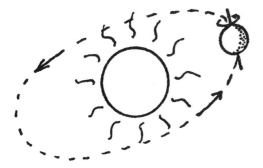

Figure 2. Spin and orbital motion.

tion, in the same way that ordinary momentum measures the "strength" of straight-line motion. For fundamental particles, one can't actually specify a rate of rotation. Nevertheless, electrons and other particles do possess measurable angular momentum.

Niels Bohr, in his landmark theory of the hydrogen atom in 1913, offered a rule of angular-momentum quantization. He suggested that Planck's constant divided by $2\pi$ (the quantity we now call *h-bar* and write $\hbar$) is the fundamental quantum unit of orbital angular momentum, meaning that an electron (and, by extension, any other particle) can possess orbital angular-momentum values only of zero, $\hbar$, $2\hbar$, $3\hbar$, and so on, not anything in between. Atomic data supported this rule until the Dutch physicists Sam Goudsmit and George Uhlenbeck discovered in 1925 that besides its orbital angular momentum, the electron also possesses spin angular momentum, of magnitude $(\tfrac{1}{2})\hbar$, only half as big as the unit formerly assumed to be indivisible. Now we know that intrinsic spin can be either half-odd-integral, as it is for electrons and quarks, or integral, as it is for photons.

A given particle has one and only one spin, a unique attribute of that particle. Or does it? Could two particles of different spin be two different forms of the *same* particle? Yes, but changing the spin is a change so drastic that we call it a different particle. It's as if you had a long wooden handle to which you could attach either an axe head or the business end of a rake. Are the axe and the rake two different manifestations of the same implement or are they two different implements? If you have the temperament of a particle physicist, you would call them different implements.

This car's driver must revere Max Planck.

For the record, $\hbar$ is equal to $1.05 \times 10^{-34}$ kg $\times$ m $\times$ m/s. The minus 34 in the exponent tells you that this spin unit is incredibly small relative to angular momenta encountered in the everyday world. How, then, do we measure such a small quantity as the spin of a single electron? There are a couple of reasons it isn't difficult. One reason is that all electrons have the same spin, so it is possible to measure the collective effect of the spins of many electrons. Another reason is that the *orientation* of spin is quantized. What this means is that a spinning particle can have its axis of rotation pointing in only a few directions, not any direction whatsoever. For an electron, with its half-unit of spin, there are in fact just two allowed directions, which we can call "up" and "down." In an atom sitting in a magnetic field, these two orientations have slightly different energies, resulting in photons of slightly different frequency being emitted by an "up" and a "down" electron. The frequency (or wavelength) can be measured to high precision, easily revealing the energy it takes to "flip" an electron spin.

The quantization of angular momentum exists in principle in the large-scale world as well. But imagine a tennis fan changing his or her "spin" from $10^{33}$ quantum units to $10^{33} + 1$ quantum units. The fractional change is far too small to notice. It's like changing the gross national product of the United States by a billionth of a billionth of a penny, or changing the mass of all living matter on Earth by less than the mass of a single bacterium.

## Units of Measurement

The units of measurement that we normally use, even in scientific work, have been defined in an arbitrary way and have nothing special to do with the fundamental laws governing the physical world. But quantum theory and relativity have revealed two "natural" units, more in harmony with the laws of nature. A still-unanswered question is whether a third natural unit awaits discovery.

The meter was originally defined as one ten-millionth of the distance from the Earth's pole to the equator. The kilogram is the mass of 0.001 cubic meter (1 liter) of water. So both the meter and the kilogram depend for their definition on the size of the Earth, and there is no reason to believe that there is anything very special about the size of the Earth. The third basic unit, the second, is also attached to a property of the Earth: its rate of rotation—again, nothing very special. For no better reason than that the Egyptians divided the day and the night into twelfths and the Sumerians liked to count in sixties, the hour is one twenty-fourth of a day, the minute a sixtieth of an hour, and the second a sixtieth of a minute.*

The two natural units that have been around now for more than a hundred years are *Planck's constant*, $h$, and the speed of light, $c$. When Max Planck introduced his constant in 1900, it was a new number perched on a completely new idea. It is the fundamental constant of quantum theory that sets the scale of the subatomic world. If $h$ were (hypothetically) smaller, atoms would be smaller and the oddities of quantum theory would be even farther removed from everyday experience. If $h$ could magically be made larger, nature would be "lumpier" and quantum phenomena would be more evident. (Don't take these remarks seriously. If $h$ were larger or smaller, everything would be different, including the scientists studying nature.)

---

* The meter and the second are now defined in terms of atomic standards, but those standards are chosen to match the original Earth-based definitions.

When Albert Einstein made the speed of light the linchpin of his relativity theory in 1905, the speed of light was neither a new number nor a new idea. But aspects of it were wholly new: that it is nature's speed limit and that it links space to time and mass to energy. So $c$ is the fundamental constant of relativity theory in the same way that $h$ is the fundamental constant of quantum theory.

Neither $h$ nor $c$ is directly a mass, a length, or a time, but each is a simple combination of these three. If they were joined by a third natural unit, they would form a basis of measurement as complete as, and much more satisfying than, the kilogram, the meter, and the second. (The thoughtful reader might propose the quantum unit of charge $e$ as a candidate for the third natural unit. Unfortunately, it won't serve, for it is not independent of $h$ and $c$, just as speed is not independent of time and distance.)

Every measurement, in whatever set of units, is ultimately the statement of a ratio. If you say that your weight is 151 pounds, you are saying, in effect, that your weight is 151 times greater than the weight of a standard object (a pint of water). A fifty-minute class is fifty times longer than the arbitrarily defined time unit, the minute. With natural units, we take the ratio with respect to some physically significant quantity rather than an arbitrarily defined one. On the natural scale, a jet-plane speed of $10^{-6}c$ is quite slow; a particle speed of $0.99c$ is quite fast. An angular momentum of $10,000\hbar$ is large; an angular momentum of $(\frac{1}{2})\hbar$ is small.

In a sense, the "natural" units $h$ and $c$ are arbitrary, too, but scientists can agree that they are directly related to fundamental features of the natural world. For an "all-natural" physics, we need one more natural unit, which has yet to emerge. This unit, if it is found, may be a length or a time. Such a unit could usher in a whole new view of space and time in the subatomic world—or, more likely, in the *sub*-subatomic world, in layers of reality far below those explored so far.

chapter 3

# Meet the Leptons

Imagine a fairy tale involving three families. The first family lives in a valley and loves vanilla ice cream. The second family lives on a mesa part way up the neighboring mountain and loves chocolate ice cream. The third family lives on the mountaintop and loves strawberry ice cream. Members of the three families have some characteristics in common and seem to be related, though anthropologists haven't figured out how, for the different families don't intermarry, and interactions among them are rare. But any member of a higher-elevation family who dies is magically transformed into three members of lower-elevation families. Members of the valley family never suffer a natural death, yet they can be wiped out by intruders from other regions. Lacking any better names, anthropologists call the three families Flavor 1, Flavor 2, and Flavor 3. That's a pretty good description of the particles called leptons. The electron and its neutrino form the valley family; the muon and its neutrino form the mesa family; and the tau lepton and its neutrino form the mountaintop family. Physicists, with no very clear idea of how the families are related, call them Flavor 1, Flavor 2, and Flavor 3. (One could say that this nomenclature is a matter of taste.)

So there are six leptons, two each of three flavors. The word "lepton" comes from a Greek word meaning "small" or "light," a reasonable term for the electron and the neutrinos. Discoveries, however, have a way of overtaking nomenclature, and, as shown in Table B.1 (in Appendix B), the tau lepton is anything but light. It's as if anthropologists had named the ice-cream-loving families the "lowland tribes" before discovering that some of them lived on a mountaintop.

Table B.1 gathers together some of the properties of the leptons. Leptons all have one-half unit of spin, and they are either neutral (uncharged) or carry one unit of negative charge. For every lepton there is an antilepton with the same mass as the lepton but opposite charge (the opposite of zero is zero). Some of the leptons are unstable. (Instability is the same thing as radioactivity. It means that after living a short—and not exactly predictable—time, the particle undergoes a sudden transformation, or *decay*, into other particles.)

What is not shown in the table is that the leptons are all weakly interacting. This sets them apart from the strongly interacting quarks (which you'll meet in the next chapter). Like quarks, leptons have no known constituents (thus, we call them fundamental) and no known size. In all experiments to date (and in successful theories describing them), the leptons act as point particles.

## The Electron

The electron, the first entry in Table B.1, is not only the best-known lepton—it has the distinction of being the first fundamental particle ever discovered.

In 1897 J. J. Thomson, a professor of physics at Cambridge University in England, wanted to learn more about cathode rays. At the time, he and other scientists knew that if a glass chamber from which most of the air had been removed contained two metal plates, and if these plates were electrically charged, one positively and one negatively, with a high voltage between them, some kind of "rays" flowed from the negative to the positive plate. Since the negative plate was called the cathode (the positive plate being the anode), the largely mysterious rays were called "cathode rays." (Names hang on. We still call the display devices common in TV sets and some computers cathode ray tubes, or CRTs.)

Thomson and his contemporaries knew that cathode rays, though normally following straight-line paths, could be deflected by magnetic

J. J. Thomson (1856–1940). Photo courtesy of Argonne National Laboratory and AIP Emilio Segrè Visual Archives.

fields. By measuring such deflections, Thomson concluded that the "rays" consisted of negatively charged particles. Moreover, using both magnetic and electric deflection,* he could measure the ratio of the mass to the charge of these particles ($m/e$). Finding that this ratio was at least a thousand times smaller than the mass-to-charge ratio of a hydrogen ion (which we now know to be simply a proton), he wrote: "The smallness of $m/e$ may be due to the smallness of $m$ or the largeness of $e$, or to a combination of these two." Based on the ability of the cathode-ray particles to pass readily through a dilute gas, he concluded, correctly, that the particles can't carry a large electric charge and so must be small in mass compared with atoms. His measurements and conclusions constitute what we call the discovery of the electron.

---

* In your present-day computer monitor or TV tube (if it is a CRT), magnetic deflection steers the electrons to various points on the screen. In a laboratory instrument called an oscilloscope, electric deflection steers the electrons.

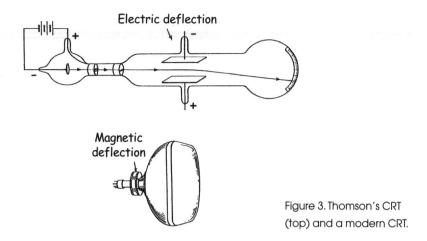

Figure 3. Thomson's CRT (top) and a modern CRT.

Thomson, in effect, launched particle physics—or subatomic physics—when he wrote that "we have in the cathode rays matter in a new state, a state in which the subdivision of matter is carried much further than in the ordinary gaseous state." Scientists realized at once that this new lightweight, negatively charged particle must be a constituent of atoms. They already knew that atoms had to be electrical in nature, for atoms could easily be ionized (that is, charged) and atoms emitted light, the result of an electromagnetic process. So the electron, once identified, fit right in. Then began the exciting process of "inventing" a structure for the atom. Niels Bohr's 1913 theoretical work on the hydrogen atom was the biggest single step in this direction, followed in the mid-1920s by the full quantum theory and its complete elucidation of atomic structure.

In 1896, shortly before Thomson's work, Henri Becquerel in France had discovered radioactivity: the spontaneous emission of "radiation" by certain heavy elements. Scientists everywhere went to work at once to find out what the radioactive atoms were emitting. Marie and Pierre Curie in France, and Ernest Rutherford, working first in Canada, then in England, were among the pioneers. These early workers learned that radioactive atoms emit three kinds of radiation. Pending knowledge of the exact nature of these radiations, they looked to the first three letters of the Greek alphabet for help and called them alpha, beta, and gamma rays. Within a few years (by 1903), alpha rays had been identified as doubly charged helium atoms (or helium nuclei, as we now know) and beta rays as electrons. So cathode rays, beta rays, and electrons are all the same thing.*

Before leaving the electron, I want to mention its antiparticle, the positron. In 1928 the English physicist Paul Dirac, as taciturn as he was brilliant,† wrote down an equation (we now call it, not surprisingly, the Dirac equation)‡ that brought together the principles of both relativity and quantum mechanics in an effort to describe the electron. To Dirac's own astonishment, not to mention the astonishment of his colleagues around the world, this equation had two startling things to say. First, it "predicted" that the electron should have spin one-half. Of course, the electron was already *known* to have spin one-half, but no one knew why, or even knew that this property of the electron would flow from a mathematical theory.

Second, Dirac's equation implied the existence of antimatter. It predicted that the electron should have a companion particle—an antielectron—with the same mass and spin as the electron but with opposite electric charge. According to Dirac's theory, when a positron and an electron met, there would be a mini-explosion. Poof! No more electron, no more positron—only a pair of photons, created in the encounter. This prediction was a bit hard for physicists to swallow at the time, for no such lightweight positive particle had ever been observed, nor had an "annihilation" event been seen. Dirac himself briefly lost faith in his own equation's prediction. He toyed with the idea that in some way the proton might be the electron's antiparticle, but soon realized that this was hardly possible. Moreover, the idea of an antiparticle with a mass different from its counterpart's was quite "inelegant."

---

* A dozen years later, gamma rays were identified as electromagnetic radiation. They were eventually recognized as photons.

† In responding to Niels Bohr's comment that Dirac rarely spoke, Ernest Rutherford told Bohr the story of a disappointed pet-store customer who returned a parrot because it didn't talk. "Oh," said the store owner, "I'm sorry. You wanted a parrot who talks. I gave you the parrot who thinks."

‡ What does a mathematical equation that is a creative achievement look like? Here it is, for a free electron:

$$(i\hbar\partial/\partial t - i\hbar c\alpha \cdot \nabla + \beta mc^2)\psi = 0.$$

Paul Dirac (1902–1984), ca. 1930. Photo copyright Cavendish Laboratory, Cambridge, England.

Dirac, like Einstein before him, believed in equations that passed the tests of simplicity, generality, and "beauty." Is this faith-based physics? Yes, in a way, but it is a faith in what sits firmly on a rock of prior knowledge—a faith that has successfully driven major advances in physics, going back to Kepler, Galileo, and Newton. Dirac argued, in effect, that his theory was too perfect to be wrong, that it was now up to the experimenters to prove it was right. And this is just what happened. In 1932 Carl Anderson at Caltech found the telltale track of an antielectron—now commonly called a positron—in a cloud chamber exposed to cosmic rays.* Before long, more confirmation was provided by Frédéric and Irène Joliot-Curie in France. They created new radioactive

---

* Carl Anderson once told me at a party that to find the age of a famous physicist, all you had to do was assume that he was twenty-six when he did his most notable work. After the party, I looked up Anderson in a reference book, and, sure enough, he was twenty-six when he discovered the positron (he turned twenty-seven in September 1932). Einstein follows this rule, and Bohr is close.

Carl Anderson (1905–1991).
Photo courtesy of AIP Emilio
Segrè Visual Archives, W. F.
Meggers Gallery of Nobel
Laureates.

elements, some of which decayed with the emission of positrons rather than electrons. Dirac, Anderson, and the Joliot-Curies were all awarded Nobel Prizes in the 1930s. (Frédéric met Irène in 1925, when he worked as an assistant to her mother, Marie Curie, herself a Nobelist. When Frédéric and Irène were married in 1926, they adopted the hyphenated name.)

We now know that every particle has an antiparticle. For a few neutral particles, the particle and antiparticle are the same. The photon, for example, is its own antiparticle. But most particles have a companion antiparticle that is distinct. All six of the leptons in Table B.1 have distinct antiparticles.

## The Electron Neutrino

Before I can introduce you to the electron's neutrino, I have to take a short detour through alpha, beta, and gamma radioactivity.

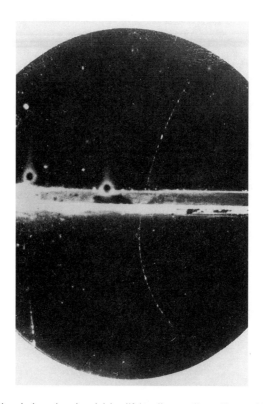

Anderson's cloud chamber track identifying the positron. He could tell that the particle flew downward because the metal plate causes the particle to lose speed, and the greater curvature of its track below the plate shows that it moved more slowly there. So the particle curved to the left, opposite to the rightward curvature that a negatively charged particle would follow. Photo by Carl Anderson; courtesy of AIP Emilio Segrè Visual Archives.

## Radioactivity

What Becquerel and his followers discovered in the closing years of the nineteenth century was that certain elements are spontaneously *active*, emitting *radiation*: hence the name "radioactivity." Within a decade or so, scientists learned that radiation doesn't ooze out gradually but is emitted in sudden "explosions" and that the energy released in each radioactive event is much greater (nearly a million times greater) than the energy released by a single atom in a chemical reaction. The sudden radioactive transformation of a nucleus* is called a *decay* event (although it has little in common with the decay of a rotting log—or an aging physicist). When a nucleus decays, it does so not according to some rigid schedule, but at an unpredictable moment limited only by probability. There is a 50-percent chance that a given nucleus will decay within a certain time span, called its *half-life*.† Either alpha or beta decay transforms it into a different nucleus, of a different element. The new nucleus may or may not be radioactive; if it is, it will have a different half-life. As mentioned earlier, what comes out of the nucleus is not "radiation" in the usual sense, but a "bullet"—an alpha, beta, or gamma particle.

That a nucleus can emit an alpha particle is not so mysterious. The alpha particle is itself a small nucleus, so one has only to visualize a small chunk of nuclear matter breaking off a larger chunk. There *is* a problem, though. Why does the alpha particle wait so long to leave its parent—sometimes millions or even billions of years? In 1928 George Gamow (a Russian émigré then living in Copenhagen), and the British physicist Ronald Gurney working with the American physicist Edward Condon, independently solved this problem. To explain the sluggishness of alpha decay, they had to use the tools of the new quantum mechanics, just then being developed. The alpha particle, they figured, could not escape its parent nucleus according to classical theory. It was held too tightly by nuclear forces. Yet according to quantum theory, it could "tunnel" free. Quantum weirdness had appeared on the scene and was here to stay. With a certain small probability, the alpha particle can

---

* For some years after the discovery of radioactivity, the inner structure of the atom remained unknown, so Becquerel and his colleagues could not pinpoint the source of radioactivity within the atom. That the *nucleus* is the seat of radioactivity became clear in 1911, when experiments by Ernest Rutherford and his colleagues in Manchester, England, established that most of the mass of the atom is contained in a tiny central core.

† Half-lives of radioactive nuclei range from thousandths of a second to billions of years. Another measure of lifetime is mean life, or average life. The half-life of a particular nucleus is 69 percent of its mean life.

Henri Becquerel (1852–1908). Photo courtesy of AIP Emilio Segrè Visual Archives.

pop through an "impenetrable" barrier and fly away, leaving behind a nucleus of less charge and less mass.

Gamma decay, once physicists realized it was the emission of high-frequency electromagnetic radiation, posed no mystery at all. Just as electrically charged electrons in the atom emit light when they jump from one quantum state of motion to another, electrically charged protons in the nucleus should be able to do the same. Because the protons oscillate with higher frequency and make higher-energy quantum jumps than the electrons, the "light" they emit is of much greater frequency than the light emitted by atomic electrons. We call such light *gamma radiation*. In modern terminology, the photons that are created (emitted) in a nuclear quantum jump are of higher energy and higher frequency than the photons created and emitted in an atomic quantum jump. In 1905 Albert Einstein had interpreted a formula advanced five years earlier by Max Planck, the formula $E = hf$, to mean that the energy $E$ carried by a photon and the frequency $f$ of the photon's electromagnetic vibration are directly proportional, the constant of proportionality being Planck's constant, $h$. So if one photon has twice the energy of another, it has twice the frequency. If a nuclear gamma-ray photon has a thousand times the energy of a photon of visible light, it also has a thousand times the frequency. Because $h$ is so small by ordinary standards, the energy $E$ of an individual photon is also small. But not zero! There is no smaller quantity of energy in radiation of a certain frequency $f$ than the energy of a single photon, given by $hf$.

So scientists in the 1920s were not particularly puzzled by gamma decay and were only a little puzzled by alpha decay (a puzzlement resolved, as mentioned, by Gamow and Condon and Gurney). But they were mightily puzzled by beta decay. Electrons shooting from the nucleus were unwelcome, for at least three reasons.

First, no one could understand how electrons could be held within the nucleus prior to their emission. Quantum mechanical theory says that an electron *can't* be confined within a nucleus. An electron held within a nucleus would have a fairly well-defined position. But because its uncertainty of position would be small, its uncertainty of momentum would be large (so says Werner Heisenberg's uncertainty principle). This means, in turn, that it would have so much kinetic energy that it would fly out of the nucleus. It's a little like trying to squeeze a balloon in your hands. The more you squeeze the balloon, the more likely it is that some part of it will swell out between your fingers, thwarting your effort to make it smaller.

Second, measurements showed that the energy carried away by electrons that are shot from a particular kind of radioactive nucleus is not always the same, and, on average, is less than the energy lost by the nucleus in the decay process. Either some energy is sneaking away in invisible, undetected form, or the sacred law of energy conservation is not sacred after all.

A third difficulty concerned the spins of nuclei before and after the beta-decay process. By measuring the spin of a nucleus, an experimenter can tell whether the nucleus contains an even or an odd number of spin-one-half particles. If a nucleus containing, say, an *odd* number of spin-one-half particles emits an electron (whose spin is one-half), what's left behind, the so-called daughter nucleus, having one less electron, should contain an *even* number of spin-one-half particles—or so it was believed. But nuclei defied this expectation. Experiments revealed that if the parent nucleus contains an odd number of spin-one-half particles, so does the daughter nucleus, and that if the count for the parent is even, it is even, too, for the daughter. One more difficulty.

## Beta Decay and the Electron Neutrino

It took two breakthroughs to unravel beta decay. The first was the suggestion by Wolfgang Pauli in Switzerland that another particle—also of spin one-half, also of small mass, but electrically neutral and unseen—was emitted along with the electron. "Dear Radioactive Ladies and Gentlemen," wrote Pauli in December 1930 to a group of physicists assembled in Tübingen, Germany, at a meeting that he couldn't attend (preferring to go to a ball in Zurich). He called his idea of the new neutral particle "a desperate way to escape from the problems" of beta decay. "At the moment, I do not dare publish anything about this idea," he added, "so I first turn trustingly to you, dear radioactive friends, with

---

* These excerpts from Pauli's letter are translated from the original German.

Wolfgang Pauli (1900–1958) at a train stop in Carlin, Nevada, on his way to Caltech in the summer of 1931. Photo courtesy of AIP Emilio Segrè Visual Archives, Goudsmit Collection.

Enrico Fermi (1901–1954), ca. 1928. Photo by G. C. Trabacchi, Ernest Orlando Lawrence Berkeley Laboratory; courtesy of AIP Emilio Segrè Visual Archives.

the question: How could such a neutron be experimentally identified?"* Needless to say, Pauli's "neutron" got a mixed reception, since no one could think of a way to observe it or measure its properties. It "solved" two of the three beta-decay problems (energy conservation and nuclear spin), but did so with an entity that seemed to be purely hypothetical.

Less than two years after Pauli's suggestion, the "real" neutron was discovered (a neutral particle about as massive as a proton and likewise found in nuclei). Following this discovery, Enrico Fermi, an astonishingly brilliant and energetic Italian physicist who would gain wider

---

* Fermi was known as a great teacher and a man of fixed habits. Every day he rose at the same hour, had the same breakfast, and listened to the same news program. As I learned in 1951 when I hiked the Jemez Mountains with him and played Parcheesi with him, he was also a man who was bent on succeeding at whatever he was doing at the moment. It was impossible not to like Fermi. He was revered by his students.

fame later as the builder of the first nuclear reactor (in Chicago),* rechristened Pauli's particle the *neutrino*, or "little neutron." The name has stuck, except that now we have to add a modifier, for, as shown in Table B.1, we know three "flavors" of neutrino. The first neutrino, that of Pauli and Fermi, is the electron neutrino.

It was Fermi who, in 1934, provided the second breakthrough that was needed to understand beta decay. He developed a theory in which a radioactive nucleus, at the moment of radioactive decay, *creates* an electron and simultaneously creates a neutrino (actually an antineutrino), both of which are instantly expelled from the nucleus. Unlike Pauli's speculation, Fermi's theory found immediate acceptance, for it solved *all* the problems that had bedeviled the understanding of beta decay. For instance, it correctly accounted for the observed spread of electron energies seen in beta decay. And, most important, it brought squarely into the middle of quantum physics an idea that has remained central ever since: that *interactions in the small-scale world occur through the creation and annihilation of particles*. This was already known to be true for electromagnetic interactions, mediated by the creation and annihilation (emission and absorption) of photons. And scientists also knew that a particle-antiparticle pair could mutually annihilate. But Fermi's creative stroke was to attribute beta radioactivity to the creation and annihilation of particles, setting the stage for this description of *all* interactions.

From then on, most physicists accepted the reality of the neutrino, even though it was not directly observed until 1956. In that year, Frederick Reines and Clyde Cowan, working at the Savannah River nuclear reactor in South Carolina, were able to catch an occasional neutrino (again, actually an antineutrino) from among the billions that the reactor sent through their apparatus every second.

In popular accounts, the neutrino is often called the "elusive neutrino," and for good reason. Neutrinos experience only the so-called *weak interaction*. Whenever a neutrino "does something"—gets emitted in beta decay, for instance, or gets absorbed in the Reines-Cowan apparatus—it is because of the weak interaction. The weak interaction is actually not the weakest interaction we know. That honor belongs to

Fred Reines (1918–1998) in his laboratory, 1950s. Photo by Ed Nano; courtesy of AIP Emilio Segrè Visual Archives.

gravity (more on this later). But it is vastly weaker than the electromagnetic interaction in which charged particles and photons participate, or the strong interaction in which protons, neutrons, and quarks participate. If a neutrino of a few million electron volts energy encountered a solid wall ten light-years thick (more than twice the distance from Earth to Alpha Centauri), it would have a better-than-even chance of getting through. This explains why detecting neutrinos is a real challenge! But it leaves you wondering: How can a neutrino *ever* be caught? The answer lies in *probability*. Think of automobiles leaving an assembly plant. Suppose that each one has a better-than-even chance of surviving 100,000 miles on the road without a serious accident. Some will go farther unscathed, some less far. A very few will go only ten miles before being involved in a serious accident, and the occasional one may survive only ten blocks. The rare neutrino detected in someone's laboratory is like a car that doesn't even make it out the factory gate without a mishap.

## The Muon

In the early 1930s, peace prevailed in the world of elementary particles. Matter was constructed of protons, neutrons, and electrons. Processes of change—interactions—involved photons and the still hypothetical but quite credible neutrinos. That was it. Five particles. This peace was not to last. In Europe and the United States, experimenters studying cosmic radiation were finding evidence for charged particles about two hundred times more massive than electrons (nine times less massive than protons and neutrons). At first, this seemed like good news. The Japanese theorist Hideki Yukawa, not long before, had postulated the existence of new strongly interacting particles of just about that mass whose exchange among neutrons and protons within the nucleus, he argued, could give rise to the strong nuclear force. A wonderful accord between theory and experiment, or so it seemed. But experiment, the final arbiter, said no. Not long after 1945, when physicists in many countries returned from war work and took up pure research again, cosmic-ray experiments showed that the Yukawa particles, though indeed real, were not the same as most of the cosmic-ray particles that had been leaving tracks in cloud chambers and photographic emulsions. Yukawa's particles came to be called *pions*. The particles more abundant in cosmic radiation were (eventually) named *muons*.*

Getting into a new field of physics is like reading a Russian novel. There are a lot of names to cope with, and at first you wonder who's

---

* The primary cosmic radiation reaching earth from outer space consists mostly of protons. Collisions of these protons with nuclei of atoms in the atmosphere give rise to all kinds of other particles, many of which don't make it to the Earth's surface. Most of the charged particles reaching the earth's surface are muons. Hold out your hand. A dozen or so muons will pass through it every second. (Also trillions—yes, trillions!—of neutrinos.)

who. So let me pause and talk about particle names. There are a lot of them, and some no longer mean what they once did.

*Leptons* are the fundamental spin-one-half particles that experience no strong interactions (and contain no quarks).

*Baryons* are composite, strongly interacting particles made of quarks and having spin one-half (or possibly $\frac{3}{2}$ or $\frac{5}{2}$). The word "baryon" comes from a Greek word meaning "large" or "heavy." The proton and the neutron and many heavier particles are baryons. But there are some heavy particles that are *not* baryons.

*Mesons*, like baryons, are composite, strongly interacting particles made of quarks. But instead of half-odd-integer spin ($\frac{1}{2}$, $\frac{3}{2}$, and so on), they have spin 0, 1, or some other integer value. "Meson" comes from a Greek word meaning "middle" or "intermediate." The word (or an alternate choice, "mesotron") was applied originally to the particles intermediate in mass between the electron and the proton. What we now call the muon was originally called the mu meson. What we now call the pion (the Yukawa particle) was originally called the pi meson. Now we know that the muon and pion have almost nothing in common. The pion is a meson; the muon is not. The pion is, in fact, the least massive meson. Many others are heavier—some much heavier, even, than the proton.

*Quarks* are the fundamental, strongly interacting particles that never appear singly. They are constituents of baryons and mesons and are themselves "baryonic," meaning that they possess a property called *baryonic charge*. When a quark and an antiquark unite to form a meson, the baryonic charge is zero. When three quarks unite to form a baryon, the baryonic charge is one. (You'll meet quarks in the next chapter.)

*Force carriers* are particles whose creation, annihilation, and exchange give rise to forces. We believe that force carriers, like the leptons and quarks, are fundamental particles with no substructure. (Force carriers likewise appear in the next chapter.)

There are other names that I'll mention here only in passing. *Hadrons* include the baryons and mesons that interact strongly; *nucleons* encompass the neutrons and protons that reside in nuclei; *fermions* (after Enrico Fermi) are particles like leptons and quarks and nucleons that have half-odd-integer spin; and *bosons* (after the Indian physicist Satyendra Nath Bose) are particles like the force carriers and mesons that have integer spin.

Back to the muon. By the late 1940s, experimenters had established that this component of cosmic radiation had no appreciable strong interaction, for it penetrated matter too readily. The muon behaved in every way like an electron, except that it was two hundred times heavier. This was a major mystery at the time, for large mass seemed to be associated only with strong interactions. Particles that did not interact strongly—electrons, neutrinos, and photons—were lightweight. Yet here was a rather heavy particle that did *not* interact strongly. "Who ordered that?" the eminent Columbia University physicist I. I. Rabi reportedly asked. At Caltech, the brilliant theorist Richard Feynman is said to have kept a curiously phrased question on his blackboard: "Why does the muon weigh?" For some time the mystery actually deepened, for every measurement that was made only confirmed that the muon and electron were, except for mass, seemingly identical.

So whatever the reason might be for the muon's existence, physicists had to accept it as a corpulent close cousin of the electron. The fact that the muon lives a mere two millionths of a second on average and the electron, if undisturbed, lives forever is not, by the way, a significant difference. Every particle would like to decay if it could. The conservation of charge, along with the conservation of energy, prevents the electron from decaying, for there are no lighter charged particles into which it can decay. The muon is not so constrained and eventually (two millionths of a second is a very long time in the subatomic realm) it does decay. The particular *way* in which it decays tells us that in fact it differs from the electron in more than mass. If mass were the only difference, we would find that the muon sometimes decays into an electron and a photon (or gamma ray), as represented by the formula

$$\mu \rightarrow e + \gamma.$$

This decay, if it occurred, would preserve lepton number (one lepton before the decay, and one after). But no such decay has been seen. Evidently the muon and electron differ in some characteristic which prevents one of them from turning into the other.

That characteristic is what came to be called *flavor*. And flavor, like electric charge and baryonic charge, is *conserved*: it's the same after an interaction as it was before. If you look at the muon-decay event listed in Table B.1, you'll see that before the decay occurs, there is one particle present with the muon's flavor—namely, in this example, the negative muon itself. After the decay, one particle with the muon's flavor, a muon neutrino, comes into being. So the muon flavor is preserved. What about electron flavor? The particles created in the decay include an electron and an *antineutrino* of the electron type. So, as physicists reckon these numbers, there are electron "flavor numbers" of $+1$ (for the electron) and $-1$ (for the antineutrino) after the decay: a total of zero, the same as before the decay. To be sure, assigning a numerical value to "flavor" seems a bit far-fetched, but it all works. Each of the three lepton flavors is conserved. (More on conservation laws in Chapter 8.)

## The Muon Neutrino

Among the experiments that established the reality of the muon in the late 1940s was one carried out by Cecil Powell and his collaborators at the University of Bristol in Great Britain. Using a special photographic emulsion as a detector, they observed the track of what we now call a pion, followed by the track of a muon, followed in turn by the track of an electron. It appeared that the pion had decayed into a muon and one or more unseen neutral particles, and that the muon had then decayed into an electron and one or more unseen neutral particles. It was likely that the unseen particles were neutrinos, but the question was: One kind of neutrino or two? Physicists had reason to believe that the muon, like the electron, would have an associated neutrino, but they could not at first tell whether the muon's neutrino was the same as the electron's or different.

Since the muon had some property other than mass that distinguished it from the electron (the property we now call flavor), it was rather natural to suppose that the muon would have its own neutrino, distinct from the electron's neutrino. But physicists like to assume that

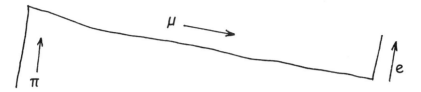

Figure 4. A drawing based on tracks left in a photographic emulsion when a pion, after being stopped, decays into a visible muon and an invisible neutrino, and the muon, after being stopped, decays into a visible electron and invisible neutrinos (actually a neutrino and an antineutrino). The original observation was made by Cecil Powell and his colleagues at the University of Bristol, ca. 1948. The length of the muon track is about 0.04 cm (less than two hundredths of an inch).

the simplest explanation is probably the correct one. Why invent a new particle if you don't need it? Perhaps the electron's neutrino could do double duty and pair up with the muon as well as with the electron. Only experiments could decide.

The critical experiment was carried out in 1962 (half a dozen years after Reines and Cowan detected the electron's antineutrino) by a Columbia University group headed by Leon Lederman,* Melvin Schwartz, and Jack Steinberger. Working at the then-preeminent 33-GeV accelerator at Brookhaven Lab on Long Island, they duplicated in the laboratory what cosmic rays do high in the atmosphere. Protons strike nuclei, creating pions. The pions fly some distance and then decay into muons and neutrinos. Given enough time and distance, the muons also decay, but these experimenters were interested in the neutrinos produced by the decaying pions. The processes were

$$\pi^+ \to \mu^+ + \nu;$$
$$\pi^- \to \mu^- + \bar{\nu}.$$

The positive pion decays into a positive muon, which is an *antimuon*, and a neutrino. The negative pion decays into a negative muon and an antineutrino. (The bar over the symbol $\nu$—Greek *nu*—desig-

---

* Lederman, who later served as director of Fermilab near Chicago, is now a national leader in the effort to reform the teaching of high school science, and has spearheaded the "Physics First" movement to teach physics in the ninth grade.

The team that discovered the muon neutrino, pictured in 1962, the year of the discovery. The team leaders are Jack Steinberger (b. 1921) on the far left, Melvin Schwartz (b. 1932) on the far right, and Leon Lederman (b. 1922) next to Schwartz. Photo courtesy of Leon Lederman.

nates the antiparticle.) To block the spray of particles coming from the accelerator, the experimenters erected a 44-foot-thick barrier of iron and concrete,* which was enough to stop just about every charged particle but posed no impediment at all to the neutrinos (and antineutrinos). If there were but a single kind of neutrino, the neutrinos that breached the barrier would have created as many electrons as muons in their detector, a chamber containing lots of protons and neutrons (within atomic nuclei!). In fact, the neutrinos created only muons,

---

* You read that correctly: 44 feet thick!

showing that the muon neutrino is *not* the same as the electron neutrino. So the reaction equations above need to be rewritten with subscripts:

$$\pi^+ \rightarrow \mu^+ + \nu_\mu;$$
$$\pi^- \rightarrow \mu^- + \bar{\nu}_\mu.$$

The Lederman-Schwartz-Steinberger group estimated that about 100 million ($10^8$) neutrinos passed through their detector each second while the experiment was running. In three hundred hours of operation, they captured twenty-nine neutrinos. The muon's neutrino is no easier to catch than the electron's!

## The Tau

So much for the leptons, thought some physicists in the 1960s. We have the electron, the muon, their neutrinos, and the antiparticles of these four leptons. That should wrap it up. Especially since, at the time, there was evidence for three quarks and theoretical reasons to believe in a fourth. (The fourth quark was discovered in 1974.) In short, physicists knew of four leptons and suspected that there were four quarks. Peace again? Not for long.

At the Stanford Linear Accelerator Center in California, Martin Perl decided to explore unknown territory, looking for a charged lepton more massive—possibly much more massive—than the muon. Perl was an activist in both politics and science. In the 1960s, he joined protests against the Vietnam War and worked for social justice at home. In the early 1970s, he led a successful struggle to create within the staid American Physical Society a unit dedicated to issues of science and society. At Stanford, some of Perl's colleagues considered his determination to look for a massive lepton unwise. He was, after all, guided only by curiosity and hope, not by theory. No theorist could tell him he would probably succeed. His experimental colleagues told him he would probably fail.

Indeed, Perl's search, culminating in the discovery of the tau, took many years. He started looking in the late 1960s and announced the

Martin Perl (b. 1927), 1981.
Photo courtesy of Martin
Perl.

first evidence for this new lepton in 1975. That first evidence was a bit
shaky. Not until 1978, after Perl and others had completed additional
experiments, did the physics community fully accept the reality of the
tau. (By that time, two more quarks had been found, bringing their total
to five.) Perl's 1995 Nobel Prize vindicated his determination and per-
sistence.

Here's how Perl found the tau. He worked with colliding beams of
electrons and positrons (antielectrons) at a "storage ring" attached to
the Stanford Linear Accelerator. In this device, electrons and positrons
are collected and "stored" (briefly!) in a large doughnut-shaped enclosure, guided in opposite directions around their underground racetrack
by magnetic fields. Each time an electron collided with a positron in
Perl's experiment, a total energy of up to 5 GeV was available, some of
which could be transformed into the mass of new particles. We are ac-
customed to thinking that when a pair of particles annihilate, their
mass energy (or "rest energy") is released. But in the Stanford Linear

Air view of the Stanford Linear Accelerator Center (SLAC). High-energy electrons and positrons are stored in underground rings in the foreground, after being accelerated down a two-mile-long pipe from the background. Photo courtesy of SLAC.

Accelerator, the *kinetic energies* of the particles vastly exceed their mass energies (which amount to only 1 MeV). When the particles annihilate, not just their mass energy but also their kinetic energy can go into making new mass. A glance at Table B.1 shows that 5 GeV is, in fact, more than enough to create a tau-antitau pair, which takes 2 × 1.777 GeV (2 × 1,777 MeV), or 3.55 GeV.

The observed reactions, interpreted as the tau's "signature," were

$$e^- + e^+ \rightarrow e^- + \mu^+ + \text{invisible particles} \tag{1}$$

and

e− + e+ → e+ + μ− + invisible particles.                    (2)

By reasoning backward from the measured properties of the electrons and muons that emerged in pairs, Perl could extract an astonishing amount of information: that between the "before" and the "after" in the reactions shown above, a massive particle and antiparticle of a new kind came into being and almost immediately decayed, and that the products labeled "invisible particles" consisted of no fewer than *four* particles—two neutrinos and two antineutrinos.

Relying on Perl's tour de force of analysis, I can now describe what

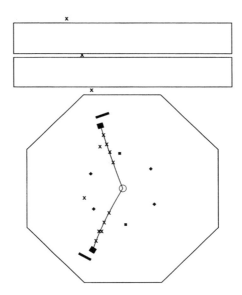

Figure 5. Computer reconstruction of an event that helped to establish the reality of the tau lepton. The small circle at the center represents the beam pipe. Electrons and positrons flew in opposite directions in the pipe—the electrons toward the viewer, the positrons away from the viewer. The track arcing downward was made by an electron, which stopped within the octagonal region. The track arcing upward was made by a muon, which continued upward, penetrating two blocks of material. Unseen but inferred were a tau-antitau pair created when an electron and positron collided within the small circle at the center. Image courtesy of Martin Perl.

happens from left to right in the reactions shown above. First, the colliding electron and positron vanish, to be replaced by a tau-antitau pair:

$$e^- + e^+ \to \tau^- + \tau^+.$$

Let's check conservation laws. Charge is zero before and after. Because of the particle-antiparticle pairs, both electron flavor and tau flavor are also zero before and after. A lot of new mass is created, but there was sufficient kinetic energy available to make that possible. The tau particles decay (swiftly!). A possible mode of decay for the negative tau is

$$\tau^- \to e^- + \nu_\tau + \bar{\nu}_e.$$

A possible mode of decay for the positive tau is

$$\tau^+ \to \mu^+ + \bar{\nu}_\tau + \nu_\mu.$$

Conservation laws enter the picture to dictate which neutrinos and antineutrinos are produced in each of these decay events. In the first one, for instance, the conservation of tau flavor requires that a tau neutrino appear to replace the vanished tau, and the conservation of electron flavor requires that an electron and an antineutrino of the electron type be created together.

If I now put together these decay modes of the $\tau^-$ and $\tau^+$ particles, I get, as a possible before-and-after representation,

$$e^- + e^+ \to e^- + \nu_\tau + \bar{\nu}_e + \mu^+ + \bar{\nu}_\tau + \nu_\mu. \tag{3}$$

This matches the reaction shown in Equation (1) above, with the invisible particles now displayed explicitly. The intermediate $\tau^-$ could equally well have decayed into a negative muon and a muon antineutrino, and the intermediate $\tau^+$ could equally well have decayed into a positron and an electron neutrino. In other words, if you change every e to a $\mu$ and every $\mu$ to an e on the right-hand side of Equation (3), you get another possible outcome. This accounts for the reaction shown in Equation (2) above.

When speculation goes too far, physicists call it "hand waving." Does the interpretation I have just given, involving *six* unobserved particles (two taus, two neutrinos, and two antineutrinos) qualify as hand waving? No, not really. The reason is that once physicists had measured the energies and the directions of travel of many muon-electron pairs generated by the electron-positron collisions, they could, remarkably, determine both the mass and the lifetime of the taus that played their invisible but crucial intermediate parts. The barest flick of time after an electron and a positron collided, an electron and a muon flew apart. What happened in the instant between these two events is something the physicists could figure out as clearly as you can figure what must have happened in the phone booth in the moment between the time Clark Kent walked into it and the time Superman walked out.

## The Tau Neutrino

The summer of 2000 brought evidence for the last lepton—believed to be the last fundamental particle that hadn't been seen before. Researchers at Fermilab in Illinois (at that time the world's most powerful accelerator) fired a beam of 800-GeV protons at a tungsten target, and then did all they could to get rid of the emerging debris, except for neutrinos. Magnets deflected charged particles out of the way. Thick absorbers captured most neutral particles. What got through the obstacle course to reach the photographic-emulsion detectors included lots of neutrinos of all kinds. A few of the tau neutrinos (about one in a million million, according to the estimates of the researchers) interacted in the emulsion to create tau leptons, which, because of their charge, left tiny tracks in the emulsion, each about a millimeter long. These tracks, along with additional tracks of particles created by the decaying taus, established the reality of the tau neutrinos.

By the time of this sighting of the tau neutrino, no physicist had any doubt of its existence. Yet all breathed a sigh of relief when their well-founded faith in the nature of the lepton family was rewarded with solid evidence. Leptons clearly have three flavors and we have a mature theory of their behavior, covering beta radioactivity and a whole host of other observed phenomena.

## Neutrino Mass

When Pauli first suggested the neutrino in 1930 (the "neutron," as he called it then), he said that it ought to have about the same mass as an electron and in any case not more than 1 percent of the mass of a proton. The idea that the neutrino might have no mass at all came a little later, as the result of beta-decay experiments. In a particular beta-decay event, there is a certain definite quantity of energy carried away by the electron and neutrino combined. The electron (the observed particle) takes some of this energy; the neutrino (the unobserved particle) takes the rest. The neutrino, if it has mass, has to take away at *least* its own rest energy, $mc^2$. Even if it dribbled out of the radioactive nucleus with no appreciable speed, it would still carry away that much energy. So the electron can take away at *most* the total available energy minus the neutrino's $mc^2$.

Physicists in the 1930s, completing successively more precise experiments, found that in beta decay the most energetic electrons shooting from the nucleus were carrying away just about all of the available energy. The mass of the neutrino, if any, had to be very small. (By now, the upper limit on the electron neutrino's mass has been pushed down to less than one hundred-thousandth the mass of an electron.) The possibility of mass exactly zero was consistent with Fermi's theory and was by then a perfectly ordinary idea, given that the massless photon was already a familiar friend. Theorists, in fact, soon developed a version of Fermi's theory that *required* neutrinos to be massless.* So for some decades it was generally assumed that neutrinos were massless particles.

A massless neutrino is an appealing idea, as simple as can be. But is it true? Apparently not. We now have good evidence that neutrinos have some small mass. Table B.1 shows current upper limits. Before looking at the evidence for neutrino mass, let's make a short detour through the subject of simplicity in physics.

### The Faith in Simplicity

Faith in simplicity has motivated scientists, and physicists in particular, for centuries. What does that mean? It is the faith that nature operates according to simple rules, and that we humans can find those rules and

---

* I once heard an experimental physicist say in a lecture that if he observed grand pianos bombarding Earth from outer space, within twenty-four hours some theorist would have developed an elegant theory showing that grand pianos are an essential component of the primary cosmic radiation.

set them down in mathematical form, usually on single sheets of paper. It is a faith that, given two possible explanations of a set of observations, the simpler is likely to be correct. What is "simple" is, of course, a matter of opinion, but scientists generally agree that it refers to conciseness, economy of concepts, brevity of mathematical expression, and breadth of application. It adds up to what scientists often call beauty.

Let me illustrate by example. Imagine a giant inflatable sphere with the Sun at its center. As the sphere is inflated, its surface area grows in exact proportion to the square of its radius. When the radius—that is, the distance from the Sun to the sphere—is doubled, the surface area grows four-fold; when that distance is tripled, the area grows nine-fold; and so on. This is a geometric fact (in three-dimensional Euclidean space). At the same time, according to Newton's law of gravitational force, the gravitational pull of the Sun weakens in exact proportion to the *inverse* square of the distance. At twice the distance, the force is four times weaker; at three times the distance; the force is nine times weaker; and so on. This is a physical law. It is hard to imagine a simpler one. The exponent on the radial distance $r$ is not 2.1 or 2.0000004; it is, according to Newton, *exactly* 2. And according to Maxwell's theory of electromagnetism, that same inverse-square law applies to the electric force between charged particles.

Yet in the twentieth century, as physicists looked deeper, they found that the marvelous simplicity of Newton and Maxwell is, after all, only approximate. Einstein's general theory of relativity shows us that the effective exponent of $r$ for gravity is *not* exactly 2 (although, for planets and stars, it is extremely close). As a consequence, an idealized single planet orbiting a star would not precisely retrace its elliptical path in every circuit (as Newton's law implies) but would take a slightly different path each time around, making for a kind of spiral path. And deep within an atom, the comings and goings of "virtual particles," a feature of quantum mechanics, make the electric force near the nucleus differ ever so slightly from the inverse-square law. One consequence of this is that two particular states of motion of the electron in the hydrogen atom that Paul Dirac in 1928 predicted should have precisely the same energy (simplicity!) turned out to differ in energy ever so slightly (not quite so simple).

One way to describe this state of affairs is to say that at a certain level of approximation—often a very good approximation—nature reveals to us laws of striking simplicity, yet shows us laws of greater complexity when we look deeper. To give one other example: according to the theories of John Wheeler and other researchers, when we look at tiny enough regions of space over short enough spans of time, the bland smoothness of space and time in our ordinary world gives way to quantum foam.

Much of the ordinary physical world around us is not simple. You need look no further than your local weather forecast to know that scientists are far from describing our physical environment in simple terms. Ripples on water, leaves trembling on trees, smoke rising from a campfire—all defy any simple description.

So, roughly speaking, we have three layers of complexity. There is, in the top layer—the visible layer—great complexity (rippling water, trembling leaves, the weather). Underlying this complexity is a layer of startling simplicity uncovered by scientists over the past few centuries (Newton's gravity, Maxwell's electromagnetism, Dirac's quantum electron). In the deepest layer, complexity rears its head again. Tiny deviations from simplicity appear. But these are not like the complexities of our immediate environment. They reflect what may be a still deeper, subtler simplicity. For example, most scientists would say that Einstein's equations of general relativity, which can be written concisely in a few lines, have a stunning simplicity (they explain gravity in terms of space and time only, and show for the first time why gravity imparts equal acceleration to all falling objects, whatever their size or composition). Yet these are the equations that tell us Newton's inverse-square law is not quite right. And modern quantum theory, with its beautiful and simple basic equations and its few concepts, nevertheless tells us that what we call a vacuum—nothingness—is in fact a lively place in which particles are constantly being annihilated and created. Simplicity is subtle, and so is the beauty that goes with it.

## Back to Neutrino Mass

So, based on what modern physics has taught us about layers of complexity and simplicity, we might say that it would be "nice" if neutrinos

Workers in a raft are inspecting the interior of the Kamiokande detector when it was partly filled with water. Photo courtesy of Kamioka Observatory, Institute for Cosmic Ray Research, University of Tokyo.

were truly massless, but that we should not be surprised if, based on some deeper—perhaps ultimately simpler—theory, they have a small mass after all. Scientists recently concluded that neutrinos do have mass, that the mass is small, and that different neutrinos have different masses.

The evidence for neutrino mass first came in the 1990s from measurements in an enormous underground detector in Japan called Super Kamiokande, and was reinforced in 2001 and 2002 by measurements at SNO, the Sudbury Neutrino Observatory (in Canada), also located deep underground.

The international group working in Japan observed muons—one every hour or so—that were created by neutrinos coursing through their apparatus. The neutrinos, in turn, had been created high in the atmosphere by cosmic rays. Because the earth is so nearly transparent to neutrinos, they saw the effects of neutrinos coming from both overhead and "underfoot"—that is, from the atmosphere on the other side of the earth—as well as from all other directions in between. What they found was that there was a "deficit" of muon neutrinos from underfoot relative to those from overhead. Some of the muon neutrinos that travel farther—some eight thousand miles farther—are getting "lost." How does this imply that neutrinos have mass? Now I must ask you to fasten your seat belts for the next few minutes.

A quantum system can exist in two "states" at once. It's further possible that a muon neutrino may not be a "pure" particle at all and may not have a single identifiable mass. It may be a mixture of two other particles, each of definite mass. The tau neutrino, in turn, may also be a mixture, but a different mixture, of the same two other particles. If the two other particles had the same mass, all of this speculation about mixing would be just so much mathematical manipulation with no observable consequences. But if the two other particles have different mass, the wave nature of particles enters the picture. The quantum waves associated with the two mixed particles oscillate at different frequency (because frequency is related to energy, and energy, in turn, is related to mass). This makes the muon neutrino turn gradually (over a flight distance of perhaps hundreds of miles) into a tau neutrino (or perhaps an electron neutrino), then back to a muon neutrino, and so on—a phenomenon called *neutrino oscillation*.

There is a musical analogy to what is going on. If two violinists in an orchestra tune up by both playing an A, but their A's are not of identical frequency (pitch), they will hear a "beat note," a slow oscillation of intensity at the *difference* of their two frequencies. When one or the other violinist retunes until their pitches are identical, the beat note disappears. In a similar way, quantum waves of two "mixed" neutrinos can "beat" to produce a slow oscillation only if the masses of the mixed particles are not identical. Neutrino mass doesn't *require* oscillation, but if oscillation is observed, it can only mean that neutrinos have mass, that the masses of at least two of the neutrinos are different, and that the particles we call muon neutrinos or electron neutrinos or tau neutrinos are themselves mixtures of other "pure mass" states.

All of this may seem like a lot to conclude just from seeing more overhead neutrinos than underfoot neutrinos. But the way in which the so-called neutrino deficit depends on angle and on energy provides compelling evidence for neutrino mass. The later work of another international team at SNO puts the case for neutrino mass on even more solid ground.

Located more than a mile and a half underground in an Ontario nickel mine, the SNO detector contains one thousand metric tons of heavy water,* and is well suited for studying neutrinos from the Sun. Based on their theoretical calculations, astrophysicists know at what rate solar neutrinos should be reaching Earth (many billions each second across each square inch), and they know, too, that all neutrinos created in the Sun are of one flavor. Only electron neutrinos can be produced by the thermonuclear reactions that power the Sun. Thanks to the pioneering work of the American physicist Ray Davis, scientists had known for many years that the number of electron-flavored neutrinos reaching the Earth is much smaller than the number expected.† The SNO results have made clear why. Because of neutrino oscillation, the electron-flavored neutrinos that leave the Sun undergo metamorphosis on their way to Earth, changing repeatedly into muon neutrinos and/or tau neutrinos, and back to electron neutrinos. The result of this *pas de*

---

* Not even a well-funded physics research group can afford to buy one thousand metric tons of heavy water. So the SNO team borrowed it, from Canada's nuclear reactor program. Someday the SNO researchers will be able to return the water, pure and for all practical purposes totally unaffected by all the neutrino bombardment it suffered.

† Some years ago, John Bahcall, an astrophysicist at the Institute for Advanced Study in Princeton, reportedly said that if the experimenters saw any fewer solar neutrinos, they would have proved that the sun is not shining.

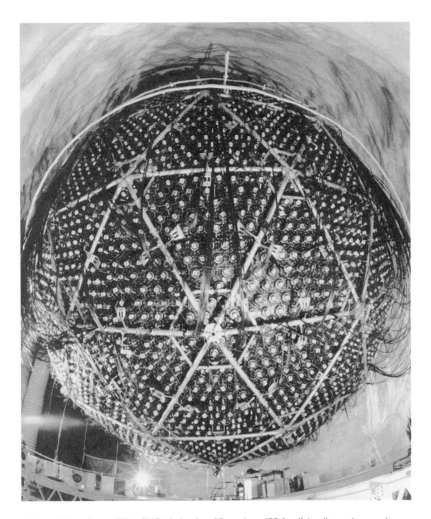

External structure of the SNO detector, 18 meters (59 feet) in diameter and holding nearly 9,600 detector tubes, pointed inward. After this photo was taken, the interior vessel was filled with a million kilograms of heavy water. Photo courtesy of SNO.

*trois* is that solar neutrinos reaching Earth are divided, perhaps about equally, among the three flavors.

To give you a flavor of the work of the SNO team, I will mention two reactions that they studied. Both involve the deuteron, which is the nucleus of heavy hydrogen. It is heavy hydrogen that makes heavy water "heavy." A deuteron consists of a proton and neutron, and I represent it by {pn}. The reactions are

$$\nu_e + \{pn\} \rightarrow e + p + p$$

and

$$\nu + \{pn\} \rightarrow \nu + p + n.$$

In the first reaction a neutron is transformed into a proton, and a neutrino into an electron. Because flavor must be preserved, the first reaction takes place only for bombarding neutrinos of the electron flavor. The researchers found only about one-third as many reactions of this type as would be expected if all of the Sun's neutrinos retained their flavor on the way to Earth. In the second reaction, a neutrino "breaks open" a deuteron, giving up some of its energy to release the proton and neutron. This reaction can proceed for any flavor of bombarding neutrino. The researchers found exactly as many reactions of this type as would be expected if all of the solar neutrinos (of requisite energy) reached Earth.

Conclusions: The Sun is emitting as many neutrinos as theory says it should (the astrophysicists can breathe a sigh of relief). Through neutrino oscillation, electron neutrinos are getting transformed into muon and tau neutrinos; and the neutrinos have masses that are not zero and are not equal.

These experiments cannot reveal exactly what the masses of the neutrinos are. We know only that they are small. Table B.1 shows current limits (as of 2003).

## Why Three Flavors? Are There More?

No one knows why there are three flavors of particles. String theorists think that their specialty might someday provide an answer. Members of this hardy band of theoretical physicists are working on the intriguing but still speculative idea that each fundamental particle consists of a vibrating or oscillating string of a size unimaginably small compared with the size of a proton or an atomic nucleus. If the string theorists are right, gravity will be incorporated into the theory of particles; the number, masses, and other properties of the fundamental particles may be predicted; and leptons and quarks will have some size after all. All of that will provide a revolution in our view of nature in the small.

If you were an explorer going where no explorer had gone before, and if you had trekked through three remote villages, you would naturally wonder whether any more villages lay ahead of you. In unknown, uncharted territory, you would have absolutely no way of knowing. You would have to march forward and find out. By analogy, you might suppose that physicists, after discovering three flavors of particles, would have no way of knowing whether a fourth flavor or a fifth or an endless number lay ahead, awaiting discovery. After all, the second flavor came as a surprise (recall Rabi's question about the muon: "Who ordered that?"), and so did the third (Martin Perl had to buck skepticism to look for the tau). Surprisingly, physicists feel confident that the third flavor marks the end of the trail—that no more lie ahead.

How can the physicist, unlike the explorer, be sure (pretty sure, at least) that the wilderness ahead holds no surprises, at least regarding more flavors? The fact that the total number of neutrinos arriving from the Sun is just three times the number of electron neutrinos that arrive is one argument that three is the limit. Another argument has to do with what in quantum theory are called virtual processes, which might roughly be called the influence of the unseen on the seen. Between "before" (say, the impending collision of two particles) and "after" (say, the emergence of newly created particles from the collision), quantum theory says that every imaginable thing happens, limited only by certain conservation laws. The intermediate things that happen include the creation and immediate annihilation of particles of all kinds. Even though these intermediate "virtual particles" are not seen, they influence the actual outcome, the "after" of the before and after. If there were one or more additional flavors of particles, they would participate in this virtual dance and their existence could be inferred. There is no evidence for such participation of particles of additional flavor.

Perhaps the most convincing argument against additional flavors comes from the decay of a heavy neutral particle called the $Z^\circ$ (pronounced "zee-naught"), which is nearly one hundred times more massive than a proton. Enter another quantum effect. The measured mass of the Z° has a range of values. Each time the mass of a Z° is measured, a different value is obtained. When the experimenters put together all of the masses they have measured for the same particle, they see a clustering of mass values about some average, with a spread that is called the *uncertainty of the mass*. (The uncertainty is meaningful because it is larger than the possible error in any single measurement. There really *is* no single mass for the particle.) According to Heisenberg's uncertainty principle, this mass uncertainty is related to a time uncertainty: more of one means less of the other. So if the mass uncertainty is big, the Z° must live only a very short time. If the mass uncertainty is small, the Z° lives longer. In fact, the lifetime of the Z° is so short that there is no way to measure it with a clock. The mass uncertainty serves as a surrogate clock. Just by finding out how the mass measurements are spread out, researchers can accurately determine the particle's lifetime.

Now comes the rest of the argument. Certain modes of Z° decay—those producing charged particles—can be observed. Among the invisible modes are decay into neutrino-antineutrino pairs. Physicists, knowing the rate of decay into visible (charged) particles and knowing the mean lifetime for all modes of decay combined, can infer the rate of decay into invisible particles. They calculate that the Z° can decay into exactly three kinds of neutrino-antineutrino pairs (the electron, muon, and tau varieties), not two or four or five or any other number. The only way that there could be more than three flavors is if the neutrinos of the fourth or later flavors were extremely massive—say, many times the mass of a proton—which is hardly expected, given that the masses of the three known neutrinos are far less than the mass of a proton.

So the flavors stop at three. But why? That is the exciting question.

## chapter 4

# The Rest of the Extended Family

## Quarks

Quarks are odd creatures indeed. They are among the limited number of particles we call fundamental, yet we have never seen one by itself—and hardly expect to. We have only a vague idea of the masses of the quarks (since we can't get hold of one to weigh it), and no idea at all why the heaviest of them is tens of thousands of times more massive than the lightest one. Yet there they are, six of them in three groups, an array that matches the array of leptons. Like the leptons, they have one-half unit of spin. And like the leptons, they have no measurable physical extension; so far as we can tell, they exist at points. Otherwise, they differ quite dramatically from the leptons, for they interact strongly (which the leptons do not), and they link up in twosomes or threesomes to make particles, such as pions, protons, and neutrons, which we *can* see (whereas leptons do not team up to make other particles).*

Imagine actors whose sole job is to get together in groups of two or three, don a horse or cow or giraffe costume, and go onstage. You, in the

---

* A positron and an electron can join forces to form a hydrogen-like atom called positronium, but that is a gigantic atom-sized entity, not a particle.

audience, can get out your digital camera, photograph the "animals," and study the photographs. You can learn a great deal about the creatures—their size and color and patterns of behavior, and (with the cooperation of your stagehand friends) their weight. You can figure out how many actors are inside each one. But you can't tell what the actors inside their animal costumes look like, or what their gender or skin color might be. You are like a physicist trying to learn about quarks by studying protons, neutrons, and other composite particles (particles made of more fundamental entities).

The stage-animal analogy is not completely apt, because in fact physicists have learned a great deal about quarks without the benefit of seeing one by itself. Moreover, there is one big difference between quarks in a proton and actors in an animal costume—namely, the measurement of their mass. If you knew that only six actors carried the full burden of playing the parts of many animals, and that they teamed up in different groups, you could figure out the weight of every actor individually by weighing every stage animal and juggling the weights until you extracted the individual weights (with a suitable allowance for the weights of the costumes). That's because in the everyday world that we inhabit, masses can simply be added. If George weighs 160 pounds and Gracie weighs 120 pounds, then George and Gracie together on a scale weigh 280 pounds. That ordinary fact is enshrined in the principle of conservation of mass so useful to chemists.

But mass, like so many things, behaves differently in the subatomic world because of the equivalence of mass and energy. When three quarks team up to form a proton, only a small part of the proton's mass comes from the masses of the quarks. Most of it comes from pure energy trapped in the proton. It's as if George and Gracie and Gloria, with a combined weight of 400 pounds, dressed up in a horse's costume and the horse weighed fifteen tons. The poor physicist, after measuring the mass of a proton to a precision of better than one part in ten million, is reduced to saying that the quarks within it have certain approximate masses, give or take a factor of two.

Table B.2 in Appendix B shows the names and some of the characteristics of the six quarks. Those in the first group have the prosaic names *up* and *down* (these are merely names, with no relation to any directions). For the second group, physicists became more whimsical, calling the quarks *strange* and *charm* (sorry, one adjective and one noun). In the late 1940s, some unexpectedly long-lived particles heavier than protons showed up in the cosmic radiation. That was "strange." These particles became the strange particles, and now we understand why they hang around for as long as a ten-billionth of a second or so. It's because they contain a kind of quark not found in neutrons and protons: the *strange quark*. Later, when some quite heavy mesons and baryons were found to live about a trillionth of a second (when they "should" have decayed much faster, in a billionth of a trillionth of a second or less), that was "charming." Needless to say, these particles' lease on life is attributable to yet another quark that they contain: the *charm quark*.

When it came to the quite heavy third group of quarks, discovered in 1977 and 1995, the physicists got cold feet. For a time, these last two quarks were called "truth" and "beauty." But conservatism prevailed. "Truth" became "top," and "beauty" became "bottom." (There was some theoretical rationale for the new names. Still, it's a pity that truth and beauty got lost.)

Although every *observed* particle carries an electric charge of zero or $+1$ or $-1$ or $+2$ or $-2$ and so on (in units of the proton charge), the quarks and antiquarks carry fractional charge: $+\frac{1}{3}$ or $-\frac{1}{3}$, $+\frac{2}{3}$ or $-\frac{2}{3}$. Yet they always combine to form observable entities with zero or unit charge.

Another property "fractionated" by the quarks is baryon number (sometimes called baryonic charge, analogous to electric charge). The proton and neutron are baryons (recall that this originally meant "heavy particle"). The critically important feature of baryon number is that it is *conserved*. A heavier baryon can decay into a lighter one. When it does so, the number of baryons before and after the decay is the same. The property called baryon number is not lost. (Similarly, electric charge is conserved. There is always as much of it after a reaction as before.) Fortunately for the structure of the universe and for us humans, the lightest baryon has nowhere to go. It is stable, because there is no lighter baryon into which it can decay. That lightest baryon is the proton. It seems to live forever.* Nearest neighbor to the proton is the neutron, just a bit heavier. This means that the neutron is unstable: it can decay into the lower-mass proton (and an electron and antineutrino) without violating the law of baryon conservation or the law of energy conservation. Left alone, the neutron lives, on average, a whole fifteen minutes before it vanishes in a puff of three other particles. Fortunately again for us humans, the neutron is stabilized within atomic nuclei, so certain combinations of up to 209 protons and neutrons can bundle together and live forever. This means that our world is made of scores of different elements, not just the single element hydrogen. And it's all because mass is energy and energy is mass. In effect, the neutron within a stable nucleus, thanks to its potential energy, is sufficiently reduced in mass that it cannot decay. (In some unstable nuclei, neutrons *can* decay. This process gives rise to beta radioactivity.)

Which brings us back to quarks. Each carries a baryon number of $\frac{1}{3}$. So three quarks, such as combine to form a proton or neutron, have baryon number 1. Since an antiquark has baryon number $-\frac{1}{3}$, a quark and antiquark together have zero baryon number. Such quark-antiquark combinations form mesons. So the antiquark is not just some curiosity. It lies within a great array of particles (none of which, however, is stable).

As you can see in Table B.2, the quarks have a wide range of masses, from a few MeV for the up and down quarks to more than 170,000 MeV for the top quark. As I mentioned earlier, no one has any idea why this is so.

Not shown in Table B.2 is another important property of quarks: *color* (sorry, one more arbitrarily chosen name—it has nothing to do with the colors that we see). Color is much like electric charge (and, in fact, is sometimes called *color charge*); it is a property carried by a particle that is never lost or destroyed. A quark can be "red," "green," or "blue." Antiquarks are antired, antigreen, or antiblue. Red, green, and

---

* Some current theories suggest that the proton may, after all, be unstable, but with such a long lifetime that the chance of its decay within the fourteen-billion-year age of the universe is miniscule. Physicists searching for proton decay have not yet found any examples.

blue combined equally are "colorless." An equal combination of antired, antigreen, and antiblue is likewise colorless.

## Composite Particles

Quarks, marvelously interesting though they are, remain hidden, as do the colors they carry. What we see in the laboratory are the colorless composite particles formed when quarks join together in twos and threes.* A few of the hundreds of known composite particles are shown in Table B.3.

The entries in the table are divided into two classes of particles: baryons, which are made of three quarks and have half-odd-integral spin ($\frac{1}{2}$, $\frac{3}{2}$, $\frac{5}{2}$, and so on), and mesons, which are made of quark-antiquark pairs and have integral spin (0, 1, 2, and so on). To pummel you with some of the vocabulary introduced in Chapter 3: the baryons are a type of particle called *fermions* and the mesons are a type of particle called *bosons*. I am saving for Chapter 7 a discussion of the quite amazing difference in the behavior of fermions and bosons. Every particle in the table is a *hadron*—that is, a strongly interacting particle—for the simple reason that the quarks of which they are formed are strongly interacting.

The lightest baryons are the proton and the neutron, which lie at the heart of every atom we know. Then come baryons with Greek names—lambda, sigma, omega (and many more). The proton and neutron are composed of u (up) and d (down) quarks. The other baryons in the table contain one or more s (strange) quarks. Still heavier baryons (not listed) contain c (charm) and b (bottom) quarks. No top-containing baryons have yet been found. The lightest charm-containing baryon is about two-and-a-half times the mass of the proton. The so-far *only* known bottom-containing baryon has a mass about six times the proton's mass.

Note that all of the baryons except the proton are unstable (radioactive). The table shows typical modes of decay and average (mean) life-

---

* In 2003, researchers in the United States and Japan reported evidence for a *pentaquark:* an entity (a baryon, actually) composed of four quarks and an antiquark.

times. The 886-second mean life of the neutron is almost "forever." Even the mean lives of about $10^{-10}$ second are extraordinarily long by subatomic standards. Recall that $10^{-9}$ is one billionth, so $10^{-10}$ second is one ten-billionth of a second. Looking down the table to the meson area, you will see that the *eta* particle has a mean life of about $10^{-19}$ second. This is unimaginably short on a human scale, yet still long enough so that the eta can sashay all the way across an atom before it succumbs.

In Table B.3 I have chosen to show only the three least massive mesons, among many dozens that are known. The lightest one of all, the *pion*, weighs in at about one-seventh the mass of a proton, yet some 270 times the mass of an electron. As mentioned in the previous chapter, when the pion was first identified in the late 1940s, it was hailed as the particle Yukawa had predicted in the 1930s—the particle whose exchange among protons and neutrons was supposed to account for the strong nuclear force. The Yukawa exchange theory is not wholly wrong, but it has been largely supplanted by the theory of gluons exchanged among quarks. So the pion, instead of being the prima ballerina in the dance of the particles, is "just" another member of the corps de ballet.

Let me explain the table's notation for the composition of the pion—or really of *three* pions, of charge $+1$, $-1$, and o. The positively charged pion is composed of an up quark and an anti–down quark. We write that combination u$\overline{\text{d}}$. If you consult Table B.2, you will discover that this combination indeed has charge $+1$ (since the charge of an anti–down quark is $+\frac{1}{3}$) and it has baryon number zero (since the baryon number of an antiquark is $-\frac{1}{3}$). All mesons share baryon number zero. The negatively charged pion is composed of a down quark and an anti–up quark, written d$\overline{\text{u}}$. The neutral, or uncharged, pion is composed of a mixture, partly an up quark and an anti–up quark, partly a down quark and an anti–down quark, so we write its composition as u$\overline{\text{u}}$ & d$\overline{\text{d}}$. Similar notation applies to the *kaons*, which also come in positive, negative, and neutral varieties. (The kaon was one of the first "strange" particles to be discovered.) The eta is, in a way, a partner to the neutral pion. Its composition is also u$\overline{\text{u}}$ & d$\overline{\text{d}}$.

Mesons can, among other possibilities, decay entirely into leptons, whereas the baryons, being constrained by the law of baryon conservation, can't do that. A baryon must always have another baryon as part of its array of decay products—although it can produce leptons, too, as the neutron does.

## Force Carriers: Particles that Make Things Happen

Here's one way to describe what physics is all about. It's about what *is* (things) and what *happens* (actions). The particles we see—leptons, baryons, and mesons—together with those we don't see but nevertheless study—quarks—constitute what *is*. There is another class of particles, called force carriers, that determine what *happens*. I should say right now that what *doesn't* happen is just as interesting as what *does* happen. There are a great many processes that don't occur (and, we think, can't occur)—the creation of electric charge from nothing, for instance, or the appearance or disappearance of energy, and perhaps the radioactive decay of a proton.

Table B.4 shows the "force carriers," particles whose exchange accounts for all the interactions, or "forces," among the other particles. They are all bosons—note the spins of 1 or 2—and all of them can be created and annihilated in any number, because no conservation law preserves the identity of any of them. Three of the force carriers have enormous mass, and three have no mass at all. Among the massless particles is the still hypothetical graviton, the force carrier of gravity.

### Gravitational Interaction

Lined up in Table B.4 are the force carriers of four different kinds of interaction, in order of increasing strength. Weakest is gravity. Because so many billions upon billions of gravitons participate in even the puniest gravitational interaction that we can measure, we see only the collective effect of gravitons in great number, never the effect of any one alone. Thus, we have no current hope of detecting the graviton. How can it be that the weakest force in nature holds us on Earth, keeps Earth in its orbit around the Sun, and can even break legs? There are two reasons. One is that gravity is attractive only, whereas the much stronger electric forces are both attractive and repulsive. Electrically, our Earth is so finely balanced in positive and negative charge that even if you get yourself charged up by shuffling across a rug on a dry day, you will feel no appreciable electric force pulling you to the ground. If all of the negative charge could be magically stripped away from the Earth, leaving only the positive charge (and if you carried about the same modest negative charge that shuffling across a rug produces), you would be instantly crushed to death by the enormous electric pull. If, conversely, all of the Earth's positive charge could be magically stripped away, leaving only the negative charge, you would be propelled electrically into space far faster than any rocket. This same fine balancing—an almost perfect cancellation of attractive and repulsive forces—prevails throughout the universe, leaving gravity as the ruling force.

The other reason that gravity is so evident to us despite its weakness is that there is a *lot* of mass pulling us down. We're held to the ground by the gravity of six thousand billion billion tons of matter. Actually each piece of matter is pulled gravitationally to every other. When we look at the gravity of ordinary-sized objects, gravity's weakness becomes plainer. When you are three feet away from the checkout clerk in a grocery store, you are pulled gravitationally toward the clerk with a force that is less than one billionth of your own weight. Put differently, the Earth's force pulling you vertically downward is more than a billion times stronger than the clerk's force pulling you laterally. No wonder it's such a challenge for physicists to measure this "lateral" force—that is, the gravitational force between two objects in a laboratory. As a result of gravity's weakness, Newton's gravitational constant, a measure of the strength of gravity, is less precisely known than other fundamental constants of physics.

Another consequence of gravity's weakness is that it plays no known role in the subatomic world. Acting between the proton and the electron in a hydrogen atom, the electric force outpulls the gravitational force by a factor that can truly be called humongous: more than $10^{39}$. (How big is $10^{39}$? That many atoms, stacked end to end, would stretch to the edge of the universe and back a thousand times.) Yet despite the incredible weakness of gravity, does it play a role at dimensions far smaller than the size of a proton? Does it become intertwined with quantum theory in ways that we can now only dimly imagine? How marvelous it will be when we find out.

## Weak Interaction

Next in the hierarchy of interactions is the weak interaction, which is responsible for the emission of electrons (beta rays) in radioactivity and for various other transformations that involve neutrinos. As its name implies, it is weak (relative to the electromagnetic and strong interactions), although it is much stronger than gravity. As shown in Table B.4, it is "mediated" (I'll explain mediation later on) by the W and Z particles, big boson bruisers, having more than eighty times the mass of a proton (the original "heavy" particle).

When Enrico Fermi developed the first theory of beta decay in 1934, he imagined direct weak interaction among a quartet of particles: the proton, neutron, electron, and neutrino. For many years thereafter, physicists speculated that one or more exchange particles (or force carriers, as we would say now) might be involved in the process, living their brief existence between the time a neutron was a neutron and the moment it vanished to become a proton, electron, and antineutrino. Not until 1983, however, did physicists discover the W and Z particles, with the aid of a large proton synchrotron at CERN in Geneva.* They are really three close siblings, with positive, negative, and zero charge, much like the triad of positive, negative, and neutral pions, which were themselves once believed to be the force carriers of the strong nuclear force. But there are a couple of big differences between the W-Z triad and the pion triad. Pions are composite particles with a significant physical size, made of quark-antiquark pairs. The W and Z particles are, we believe, fundamental, lacking any physical size and not made of anything smaller. Moreover, the W and Z particles are enormously massive relative to pions.

---

* CERN (pronounced "surn") is the European Center for Nuclear Research, with its letters arranged to match the order of the words in French, Centre Européen pour la Recherche Nucléaire. CERN's synchrotron, like others around the world, is a racetrack accelerator in which synchronized pulses of electric force are applied to the particles as they speed past designated points on the track.

### Electromagnetic Interaction

The photon, the next entry in Table B.4, has had an interesting history, from its "invention" by Albert Einstein in 1905 through its somewhat shadowy existence in the 1920s as a "corpuscle," not quite a real particle; on through its central role in the 1930s and 1940s, when physicists linked it with the electron and positron to build the powerful theory called quantum electrodynamics; up to the present day, when we see it as a fundamental particle of truly zero mass and truly zero size serving as the force carrier for electromagnetism. Indeed, you literally "see" photons nearly every waking moment, day and night. They carry part of the Sun's energy to Earth, and bring the light emitted by every star, planet, candle, light bulb, and lightning flash to your eyes. Billions of photons each second carry information from the page you are reading. And there are lots of photons that you don't see—those that carry radio and television and wireless-phone signals, heat from warm walls, and X rays through your body. The universe is filled with low-energy photons, the so-called cosmic background radiation, left over from the Big Bang. All in all, there are about a billion photons in the universe for every material particle.

I haven't yet mentioned something that Table B.4 hides: a wonderful unification of two of the four kinds of force—weak and electromagnetic. Yukawa realized long ago that the more massive an exchange particle, the less its reach, the less the "range" of the force. If you imagine yourself as a god who can change the mass of an exchange particle to suit your whim, you find that as you make that particle more and more massive, leaving everything else unchanged, the force gets weaker and weaker (in addition to having shorter and shorter range). In the 1970s, three eminent theoretical physicists—Abdus Salam, Steven Weinberg, and Sheldon Glashow—boldly advanced the idea that the weak and electromagnetic interactions are different faces of a single underlying interaction. In effect, they said that the essential difference between these two kinds of interaction is only the difference in the nature of the force carriers. The electric force is of long range (we can sense it over meters or even miles) and is relatively strong because the carrier of this

Abdus Salam (1926–1996), ca. 1978. Photo courtesy of AIP Emilio Segrè Visual Archives, *Physics Today* and Weber Collections, and W. F. Meggers Gallery of Nobel Laureates.

force is a particle with no mass, the photon. The weak interaction is of short range (reaching a distance smaller than the diameter of a proton) and is relatively weak, so it requires very massive force carriers. Not many years later, the discovery of the W and Z particles provided the needed confirmation of this "electroweak" theory.

In some sense, then, the weak and electromagnetic interactions are one and the same force. Not quite, however, for the weak interaction is universal, affecting all particles of all kinds. The electromagnetic interaction affects only particles with electric charge.

The physicists who achieved this unification shared a Nobel Prize for their work in 1979. Salam, an urbane Pakistani, headed the theoretical physics department at Imperial College in London (where I had the privilege of working in his group in 1961–62). He also led the successful

Steven Weinberg (b. 1933), 1977. Photo courtesy of AIP Emilio Segrè Visual Archives, Weber Collection.

effort to create the International Centre for Theoretical Physics, in Trieste. From that base, he worked tirelessly to assist and encourage struggling physicists in the less developed parts of the world.

Weinberg was at Harvard when he worked on the electroweak theory and moved later to the University of Texas at Austin.* His contributions to theoretical physics span an enormous range. In addition, he is a graceful and effective writer on physics for the general reader. Weinberg and Glashow were fellow high school students in New York

---

* According to legend among physicists, Weinberg made an "outrageous" salary demand when he moved to Texas: he asked to be paid as much as the football coach. I don't know if the story is true or, if it is true, whether the university met his demand.

Sheldon Glashow (b. 1932), 1980. Photo courtesy of AIP Emilio Segrè Visual Archives, Segrè Collection.

City and later Harvard professors together. Glashow, the son of Russian Jewish immigrants who had no college education, became not only a major contributor to particle physics but also a popular and effective teacher of nonscience students.

## Strong Interaction

The last entry in Table B.4 is a set of eight particles (sixteen if you count the antiparticles). These are the aptly named gluons, which provide the strong-interaction "glue." Although not electrically charged, they carry odd mixtures of color charge, such as red-antigreen or blue-antired. There are eight independent combinations of color-anticolor that define the eight gluons.* Every time a quark interacts with a gluon, the quark's color changes. It's as if every time you paused at a stand selling T-shirts (or "interacted" with a T-shirt stand), you traded in the T-shirt you were wearing for one of a different color. Unlike photons, which

---

* Red, green, and blue are now the standard designations of the strong-interaction "colors." There was a time when physicists from different countries preferred to use the colors in their national flags.

can interact with one another only indirectly via charged particles, gluons exert a direct force on one another, in addition to exerting forces on quarks.

Why do physicists come up with unseen things like quarks, gluons, and color? Because these things work. Six quarks, eight gluons, three colors—the array begins to sound like a kaleidoscope held together by duct tape. But the number of facts accounted for by the scheme far exceeds six plus eight plus three. Hundreds, if not thousands, of particles and their mutual interactions are brought to order by this "picture" of the strong interactions.

To visualize gluons, insofar as that's possible, think of a single proton blown up to the size of a basketball. Now remove the "skin" of the basketball, leaving a spherical chunk of space. Milling about in that space, never straying out of it, are three quarks. No harm if you want to think of them as gaily painted in three different colors—except for one detail: quarks are points. They are entities that have no size whatsoever, yet possess mass, color, spin, electric charge, and baryon number. As quarks flit about, they are incessantly emitting and absorbing gluons. So the space that was the inside of the basketball contains not three but dozens of particles, all in a mad dance of motion, creation, and annihilation. The gluons, like the quarks, are points—points with color, anticolor, and other properties, but without mass. Miraculously, or so it seems, the swarm as a whole maintains color neutrality. It also maintains a total electric charge of one unit (this is a proton), a total baryon number of one unit, and a total spin of one-half unit.

If one quark starts to stray outside the boundaries of the original basketball surface, it gets tugged strongly back by the gluons, in the way that a child who starts to wander away from a designated play area might get tugged back by a vigilant teacher. This strong force is quite remarkable. Unlike gravity (which gets weaker with increasing distance) or the electric force (which also gets weaker with increasing distance), the pull of gluons *increases* as the distance gets greater. So the space occupied by the quarks and gluons, the space that is the interior of a proton, needs no "skin." The gluons police the quarks and one another relentlessly at the boundaries, making sure through stronger and stronger force that no particle wanders away. There is evidence that right in the middle of the proton, each quark meanders relatively freely (like a child in the middle of the play area who can safely be ignored by the teacher).

So we have never detected a quark or gluon by itself. Since the strength of the force increases as distance grows, it is impossible to tear one of the particles loose from the others. Yet if you draw an analogy to a rubber band, which pulls more strongly the more it is stretched, you might wonder if it isn't possible to pour so much energy into a proton that even the strong gluon bond snaps, popping a quark free, just as a rubber band, with a sufficiently energetic pull, will break and release whatever is attached to its ends. As so often happens, the analogy from the ordinary world around us doesn't translate well to the subatomic world. Once again, the equivalence of mass and energy makes itself felt. Yes, with sufficient energy poured into a proton (by slamming it, let's say, with another proton fired from an accelerator), a gluon-quark bond can be broken. But the gluons and quarks have a way of fighting back against this disruption. The energy that is required to cut loose a quark is enough to make assorted other particles. If, for instance, you have succeeded in snipping free a quark with a great burst of energy, some of that energy will transform itself into other quarks and antiquarks. One of the new antiquarks will embrace the quark that was about to break free, and what you will see in your measuring apparatus will be not a free quark but a pion. You freed a quark from the proton, but it left with an antiquark chaperone, thwarting your desire to see it alone.

## Feynman Diagrams

The American physicist Richard Feynman (pronounced "FINE-mun") —celebrated for his wit and his writing as much as for his brilliant contributions to physics*—invented a method of diagramming events in the subatomic world that is a great aid to visualizing what is going on there. In particular, these "Feynman diagrams" reveal what we think is

---

* Once, when I had the privilege of giving a talk at Caltech, Feynman sat in the front row, apparently asleep. When I finished, he raised his hand and asked a question so penetrating that I was taken aback. I had to go home and work hard to find an answer.

Richard Feynman (1918–1988), Los Alamos badge photo, ca. 1943. Photo courtesy of Los Alamos National Laboratory and AIP Emilio Segrè Visual Archives.

"really" happening when a force carrier gets exchanged between two other particles, accounting for an interaction between them. In the hands of a theoretical physicist, Feynman diagrams amount to more than an aid to visualization. They provide a way to catalog possible re-actions among particles and even to calculate the likelihood of various reactions. Here, however, I want to use them just as a visual aid, so that you can see more clearly how the exchange forces produced by the force carriers work.

A Feynman diagram is a miniature spacetime map. To sneak up on this idea, let's start with a plain ordinary map, a "space map," such as might be found in a road atlas. Typically, such a map is two-dimensional, with north pointing up and east pointing to the right. A line on the map traces a path through space—or, as a mathematician would say, a projection onto the ground of a path through space. In Figure 6, lines show the paths of an airplane flying due east from Chicago's Midway Airport to Toledo; an automobile traveling in a southwesterly direction from Toledo to Indianapolis; and a proton circling in the Tevatron at Fermilab (its path much enlarged). Arrows on the lines tell the direc-

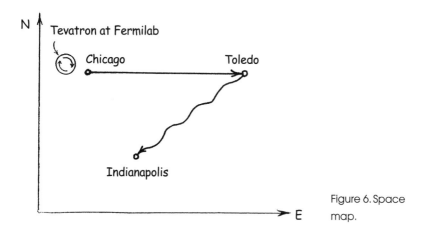

Figure 6. Space map.

tion in which the motion occurred. If we wanted to know the altitude at which the airplane flew or the distance underground of the circling proton, we would need to extend the map into a third dimension, showing "up-down." Even then we couldn't tell from the map *when* the airplane or the automobile or the proton was at a particular point. The lines on the map show only the paths through space. To include time as well as space, we would need to go to four dimensions—that is, into the realm of spacetime, which gets beyond our powers of visualization. (Physicists well versed in relativity are no better than the average person in visualizing four dimensions.)

Fortunately, a two-dimensional version of the spacetime map is quite useful, just as a two-dimensional road atlas is useful. Such an abbreviated spacetime map is shown in Figure 7. The horizontal axis, labeled x, measures distance along an east-west line. The vertical axis, labeled t, measures time. If you are standing still, your "path" on an ordinary space map is a point. You are just "there." Not so on a spacetime map, because we move inexorably forward through time. So the airplane sitting on the runway in Chicago prior to takeoff traces a vertical line (AB in the diagram). Position in space is fixed, but position in time is not. Then the airplane takes off and flies at approximately constant speed to Toledo. Can you see that its spacetime path, BC in the diagram, is a line sloping upward to the right? The line goes to the right because the airplane is moving in that direction. The line moves up be-

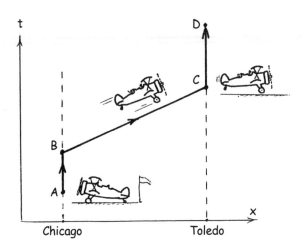

Figure 7.
Spacetime map.

cause time is passing. If the airplane flew slower, its path would be more nearly vertical (closer to the vertical line that represented no motion at all). If the airplane flew faster, its path would be more nearly horizontal (covering the same distance in less time). Finally, it lands in Toledo and parks on the ramp. Then its spacetime path is again a vertical line (CD in the diagram).

The airplane's spacetime path ABCD is called its *world line*. Note that that line is not moving. It's just "there," like a line on an ordinary map. It is the history of the trip in both space and time. (Of course, you could imagine that the world line is "evolving," being known at any time only up to that time. In the same way, if you traced your path on a space map as you traveled, you could trace it only from your starting point to your present location, not being exactly sure where the trace might go in the future.) So a spacetime map tells us about both the where and the when of a "thing"—an airplane, an automobile, or a subatomic particle. For an airplane or an automobile, there might be more going on that the world line doesn't fully reveal—a flight attendant serving lunch, or children playing in the back seat. But in the case of a fundamental particle, there is no hidden activity. Each segment of the world line tells all there is to know. But when one segment gives way to another segment, at a point where an interaction occurs, that's a different story.

In relativity theory, an "event" is something that happens at a particular spacetime point—that is, at a point in space and at an instant of time. Points B and C in Figure 7 represent (approximately) events. At B, something happens: The airplane starts moving and takes off. At C, something happens: The airplane lands and rolls to a stop. Neither of these airplane "events," of course, happens at a single point in space or a single instant of time, but they illustrate the idea. In the particle world, events seem to occur at exact spacetime points, not spread over space and not spread over time. Indeed, experiments indicate that *everything* that happens in the subatomic world happens ultimately because of little explosive events at spacetime points—events, moreover, in which nothing survives. What comes into the point is different from what leaves it.

Before turning to some spacetime diagrams for particles, let's look at one more feature of the airplane's world line. In Figure 7 it is shown with arrows. Isn't this redundant? After all, there is only one way for the airplane to "move" between points A and B—upward, because that is the direction in which time advances. Only in science fiction could an airplane move backward in time. So why bother to show the arrows? Because when we translate these ideas to particles, we need the arrows. Subatomic particles live in a world ruled by a science that sometimes seems like science fiction. Particles can, in fact, move backward in time, and do so continually as they execute their interaction ballets. It was Feynman's teacher John Wheeler who conceived this idea of particles moving backward in time; Feynman incorporated the idea into the diagrams that bear his name.

Figures 8–12 show some sample Feynman diagrams for particle processes. In each sketch, read time from bottom to top. Imagine placing a ruler horizontally near the bottom of a diagram and moving it slowly upward to mark the passage of time. In Figure 8, for instance, you see two electrons approaching each other, and then, after their interaction, two electrons receding from each other. This is called electron *scattering*, represented by

$$e^- + e^- \rightarrow e^- + e^-.$$

The diagram shows only the simplest among myriad possible sequences, but it is one that we are pretty sure really happens. At point A,

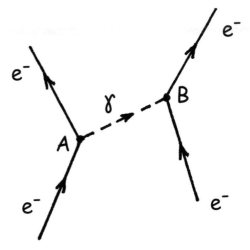

Figure 8. Electron-electron scattering.

one electron emits a photon (a gamma ray) and at point B, this photon is absorbed by the other electron. A photon has been exchanged, and as a result the electrons change speed and direction. This is the electromagnetic interaction at work.

There are two important features of this diagram that show up in all Feynman diagrams; one is obvious, and one is not so obvious. The obvious one is that at the interaction points A and B, three particle lines meet. A point such as A or B is called a *vertex*, specifically a three-prong vertex. It is the spacetime point where the interaction occurs. If you glance at the other diagrams, you'll see that all of them have three-prong vertices. Moreover, they are vertices of a particular type, where two fermion lines and one boson line meet. The fermions in these diagrams are either leptons or quarks, and the bosons are force carriers—photons, W bosons, or gluons. Here is the stunning generality that physicists now believe to be true. Every interaction in the world results ultimately from the emission and absorption of bosons (the force carriers) by leptons and quarks at spacetime points. Three-prong vertices lie at the heart of every interaction.

The not-so-obvious feature in Figure 8 is that the interaction event is a truly catastrophic event in which every particle is either annihilated or created. At point A, the incoming electron is destroyed, a photon is

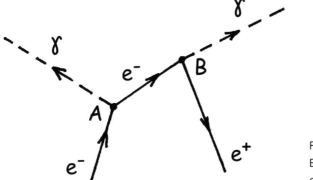

Figure 9.
Electron-positron
annihilation.

created, and a *new* electron is created. The electron flying upward to the
left in the figure cannot be said to be the same as the electron that entered from the lower left. They are *identical*, because they are both electrons, but saying that the one leaving is the *same* as the one that arrived
has no meaning. Elsewhere in Figures 9–12 the distinctness of the three
particles that converge at a vertex is more evident.

Figure 9 shows one way in which an electron and its antiparticle, a
positron, can meet and annihilate to create two photons, a process represented by

$$e^- + e^+ \rightarrow 2\gamma.$$

Once again there are two interaction vertices, A and B, where two
fermion lines and one boson line meet. At A, an incoming electron
emits a photon and creates a new electron, which flies to B, where it en-
counters an incoming positron and emits another photon. To think of
this process proceeding forward in time, you could again imagine a horizontal ruler moved slowly upward, and *ignore the arrows*. You may reasonably ask: Why are the arrows there if I'm being asked to ignore them?
Because they are labels. Their purpose is to tell whether the line is that
of a particle or an antiparticle. So the line on the right with a down-
ward-pointing arrow represents a positron moving forward in time—
that is, upward in the diagram. But (and here is where the Wheeler-
Feynman vision kicks in) the forward-in-time positron is equivalent to a

backward-in-time electron. So it is also possible to interpret this diagram more in the way it appears to the eye. An electron comes along from the left, moving forward in time, emits photons at A and B, and then reverses its course through time. Strange but true. Wheeler and Feynman showed that the descriptions in terms of a forward-in-time positron and a backward-in-time electron are both "correct" because they are mathematically equivalent and indistinguishable. Yet—you might protest—you and I don't have the option of moving forward or backward in time. We move inexorably forward, like the horizontal ruler that slides upward over the diagram. What we *see* in Figure 9 is a positron moving to the left on its way to a collision with an electron, even though, instructed in the ways of the quantum world, we are prepared to believe that what we are seeing can also be described as an electron backpedaling through time and moving to the right as time unwinds.

An example of the weak interaction at work appears in Figure 10—a Feynman diagram that displays the decay of a negative muon,

$$\mu^- \rightarrow e^- + \nu_\mu + \bar{\nu}_e.$$

Here the $W^-$ boson plays the intermediate role as an exchange particle. At each vertex, you can see conservation laws at work. At point A, one unit of negative charge arrives and one unit of negative charge leaves; one particle of the muon flavor arrives (the muon itself) and one leaves (the muon neutrino). At point B, charge is again conserved, and electron flavor is preserved (zero before and zero after) because the created electron and the created antineutrino have electron-flavor numbers $+1$ and $-1$, respectively. As the arrows suggest, and as we can infer from the discussion just above, we can also look at vertex B as a point where a backward-in-time neutrino absorbs a $W^-$ particle to become a forward-in-time electron.

The muon-decay process shown in Figure 10 can actually be regarded as a beta-decay process because an electron is created and emitted, just as in the beta decay of a radioactive nucleus. Figure 11 depicts the similar process for neutron decay:

$$n \rightarrow p + e^- + \bar{\nu}_e.$$

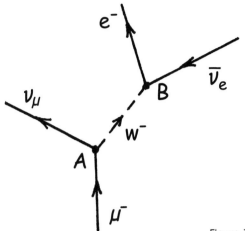

Figure 10. Negative muon decay.

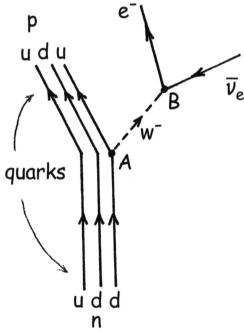

Figure 11. Neutron decay.

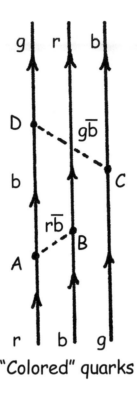

"Colored" quarks

Figure 12. Quarks exchanging gluons.

You can see that Figures 10 and 11 are very much alike. In one case, a negative muon is transformed, via the weak interaction, into a muon neutrino. In the other case, a down quark is transformed, also via the weak interaction, into an up quark. The strong interaction plays no role in the decay event, but it holds the three quarks together in the initial neutron and the final proton.

Finally, Figure 12 shows, very schematically, how gluon exchange works among three quarks (which could constitute either a proton or a neutron). This time, we label the three quarks by letters representing their color-charge: r for red, g for green, and b for blue. The "upness" or "downness" of a line doesn't change as the interactions proceed. So the quark represented by the left-hand line, which cycles through red, blue, and green, remains always an up quark or always a down quark. Four sample interaction vertices are shown (note that each of them is a three-prong vertex where two fermion lines and one boson line meet). At A, the red quark emits a red-antiblue gluon (designated r$\bar{\text{b}}$), becoming a blue quark. Later, at D, this blue quark absorbs a green-antiblue gluon to become a green quark. At B, a blue quark absorbs the red-antiblue gluon that was emitted at A to become a red quark. And at C, a green quark emits a green-antiblue gluon to become a blue quark. Stretch your mind to imagine this dance of emission, absorption, and exchange of color happening billions of times each second within a single proton. Because of the nature of the strong force, pulling ever harder as the particles get farther apart, the quarks, try as they might, are unable to get free. They continue to dance their pas de trois forever.

## chapter 5

# Quantum Lumps

Max Planck did not set out to be a revolutionary. When he presented his theory of radiation to the Prussian Academy in Berlin in December 1900 and introduced his now-famous constant $h$, he thought he was offering a refinement to classical theory, fixing a little flaw in a solid edifice. (When the quantum revolution he triggered gained momentum in the years that followed, Planck wanted no part of it. He couldn't embrace what he had initiated.)

Planck was fixing a problem that showed up when the theories of electromagnetism and thermodynamics were blended. Electromagnetism deals with light (and other radiation), as well as with electricity and magnetism. Thermodynamics deals with temperature and the flow and distribution of energy in complex systems. These two theories, pillars of nineteenth-century physics, were not up to the task of explaining "cavity radiation"—the radiation inside a closed container at a fixed temperature.

As Planck and his contemporaries knew, any object, at any temperature, emits radiation. The higher the temperature, the greater the *intensity* of the emitted radiation and the greater the average *frequency* of the radiation. These rules sound a little complicated, but ordinary experi-

Max Planck (1858–1947). Photo by R. Dührkoop; courtesy of AIP Emilio Segrè Visual Archives, W. F. Meggers Gallery of Nobel Laureates.

ence confirms them. The heating element of an electric stove, at a low-temperature setting, emits some radiation, mainly in the infrared range of frequencies, which you sense as heat if you hold your hand above the element. At a high-temperature setting, the heating element radiates more intensely and some of the radiation has moved from lower-frequency infrared to a higher-frequency red glow of visible light. The idea of radiation by a cold body is less familiar, but even the ice cap at the

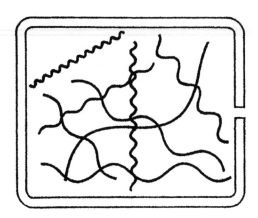

Figure 13. Different
frequencies within a cavity.

North Pole radiates. It does so much less intensely than a heating element and at lower frequencies, yet sufficiently to send some of the
Earth's energy back into space.

Inside a "cavity" (your living room, for instance) radiation of many
frequencies is bouncing around, being emitted and absorbed by the interior walls. Prior to Planck's work, scientists had learned a remarkable
fact about this radiation: its properties depend only on the temperature
of the walls, not on the material of which they are composed. Yet despite this marvelous simplification, all efforts to account for the way in
which the intensity is distributed over different frequencies failed.

In October 1900, Planck had obtained a formula for the distribution
of radiant energy in a cavity that fit the facts remarkably well. But his
formula had no theoretical basis. It was what physicists call "curve-
fitting." He felt that his formula, which worked so well, must be right—
but why did it work? What was going on, at the level where atoms and
molecules interact with radiation, that would explain it? He set to work.
"After a few weeks of the most strenuous labor of my life," he said later,
"the darkness lifted and a new, unimagined prospect began to dawn."
He found he could account for the observed features of cavity radiation
by postulating that a vibrating charge emits radiation not continuously
like water from a fire hose but instead in lumps like baseballs from a
pitching machine. He called these lumps "quanta." Thus was born the
quantum theory.

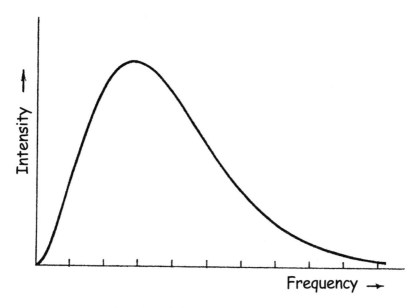

Figure 14. Intensity of cavity radiation.

Specifically, Planck had to postulate that the minimum energy quantum a vibrating charge can emit is directly proportional to the frequency of vibration; so twice the frequency means twice the minimum energy, three times the frequency means three times the minimum energy, and so on. A direct proportionality can be expressed in a simple equation with a constant of proportionality. The formula is

$E = hf,$

where $E$ is the quantum energy, $f$ is the frequency, and $h$ is the constant of proportionality, which we now call Planck's constant. Planck could deduce the numerical value of $h$ by fitting his formula to the available data. He obtained a number that differed by only a few percent from the accurate number we know today.

You encountered this formula in Chapter 3 as an expression for the energy of a photon. That is the interpretation Einstein gave in 1905. Five years earlier, in Planck's hands, the formula had been used only to express the way in which radiant energy is emitted by a vibrating

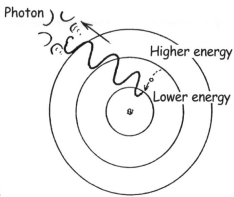

Figure 15. Energy change
determines photon frequency.

charge. Planck made no suggestion that the radiant energy is itself
quantized (that is, lumpy); he said only that energy is added in quantized amounts. It's as if you could add to a swimming pool one bucket of
water, or two, or three, or any number of full buckets, but not any quantity less than one bucket. Yet the water in the pool could be divided up
any way you like, not just by the bucketful (or so went the pre-photon
reasoning).

Planck embraced without question the classical prediction that the
frequency of the emitted radiation is the same as the frequency of vibration of the charge that is doing the emitting—just as the frequency of
sound emitted by a banjo is the same as the frequency at which the
banjo string is vibrating. (Einstein made this same assumption with his
photons.) Not until Niels Bohr applied quantum ideas to the hydrogen
atom in 1913 did it become clear that what determines the frequency of
emitted radiation is the *energy change* in the emitting system. According
to Bohr, an electron could be vibrating at one frequency before it emitted light and another frequency after it emitted light, with the frequency of the emitted light being neither of those electron frequencies.
Planck's formula survived this changed outlook, but again with a different interpretation. In Bohr's hands, the $E$ in $E = hf$ is the energy change
of the material system and the $f$ is the frequency of the emitted radiation, *not* the frequency of material vibration. (Bohr, in 1913, had not
quite accepted Einstein's photon, so he didn't discuss it. But we would
now say that the loss of energy $E$ of the material system is equal to the
energy gained by the created photon.)

The white interior of the box is black when viewed through the hole in the side because nearly all light that enters is absorbed and doesn't get out again. The box contains cavity radiation. Photo by Robert Douglas Carey; courtesy of Pearson Education and Helen Yan.

Cavity radiation is often called *black-body radiation*. An ideal "black body" is one that absorbs all radiation striking it. Such a body radiates energy with the same intensity and distribution of frequencies that characterize cavity radiation. There is a simple reason for this. Radiation bouncing around within a cavity is eventually fully absorbed, so the interior of a cavity is "black" (even if it's painted white!). The surface of the Sun is also very close to being a black body—even though it, too, is far from black in color. When John Wheeler coined the name "black hole" in 1968 for a fully collapsed state of matter, he had in mind its analogy to a black body. A black body absorbs all radiation that hits it, but also radiates energy. A black hole absorbs everything that hits it (radiation and matter both), but emits nothing back.*

When Planck introduced the quantum idea in 1900, he was forty-two, well beyond the age at which many theoretical physicists do their most notable work (as I mentioned in Chapter 3, Carl Anderson put this age at twenty-six). Among the physicists who developed the full theory of quantum mechanics a quarter-century later, in the period 1924–1928, some hadn't even been born when Planck launched the revolution. In 1900, Max Born, Niels Bohr, and Erwin Schrödinger were teenagers, Satyendra Nath Bose was six, and Wolfgang Pauli was an infant. Werner Heisenberg and Enrico Fermi came into the world in 1901, and Paul Dirac in 1902. We often say that quantum mechanics (as a theory of the subatomic world) was "created" or "completed" by these mostly young people in the mid-1920s. In a sense that is true, but mysteries remain. Many physicists still regard quantum mechanics as a work in progress. It's not that quantum mechanics makes any wrong predictions, or that it hasn't been able to cope with particles like quarks, concepts like color, or trillion-volt energies. It's just that the theory seems to lack a rationale. "How come the quantum?" John Wheeler likes to ask. "If your head doesn't swim when you think about the quantum," Niels Bohr reportedly said, "you haven't understood it." And Richard Feynman, the brash and brilliant American physicist who understood quantum mechanics as deeply as anyone, wrote: "My physics

---

* Well, almost nothing. The British theoretical physicist Stephen Hawking discovered that black holes do, after all, radiate slightly. The reason? A subtle quantum effect. See Chapter 10.

students don't understand it either. That is because *I* don't understand it."* Many physicists believe that some *reason* for quantum mechanics awaits discovery.

One thing that has stood the test of time since that day in Berlin in December 1900 is Planck's constant. It remains the fundamental constant of quantum theory, with ramifications that go far beyond its original role in relating radiated energy to radiated frequency. As I noted earlier, it is the constant that sets the scale of the subatomic world and that distinguishes the subatomic world from the "classical" world of everyday experience.

This chapter is about quantum lumps, of which I have discussed only one so far: the lump (or quantum) of radiated energy that becomes the lump (or particle) of radiant energy known as the photon.

There are two kinds of lumpiness in nature: the granularity of *things* and the granularity (discreteness) of certain *properties* of those things. Let's start with the granularity of things. Everyone knows that you can't subdivide a piece of matter indefinitely. If you divide it finely enough, you get to atoms (the word "atom" was originally chosen to mean "indivisible"), and if you pull atoms apart, you get to electrons and atomic nuclei and eventually to quarks and gluons. So far as we know, that's as far as you can go. We know of no size and no structure for electrons and quarks. "Well," you might ask, "isn't that only because we haven't yet learned to probe more deeply? Why shouldn't there be worlds within worlds within worlds?" Scientists have a couple of reasons for believing that the onion of reality has a core with only so many layers to uncover, and that we have reached, or are very close to reaching, that core.

One reason for this conclusion is that it takes only a few quantities to completely describe a fundamental particle. An electron, for instance, is described by its mass, charge, flavor, and spin, and by the strength of its interaction with force-carrying bosons of the weak interaction—and that's about it. Physicists are confident that if there are any properties of the electron still to be discovered, these properties are few in number. So it takes only a short list to specify everything there is to specify about an electron. Contrast that with what it would take to

---

* Richard Feynman, *QED* (Princeton, N.J.: Princeton University Press, 1985), p. 9.

completely describe a "simple" steel ball bearing. Normally we would say that we know all we need to know about the ball bearing if we know its mass, radius, density, elasticity, surface friction, and perhaps a few other things. But those properties don't begin to describe the ball bearing completely. To describe it *fully*, we would need to know how many atoms of iron it contains, how many of carbon, how many of various other elements, how these atoms are arranged, how the countless electrons are spread throughout the material, what energies of vibration are jiggling the atoms, and more. You can see that a list describing the ball bearing down to the last detail would have billions upon billions upon billions of entries. The description of matter does seem to get simpler as we peel away the layers of the material onion, and apparently can't get very much simpler than it is for the fundamental particles.

Another reason for believing that we are close to a genuine core of matter—a reason closely related to simplicity of description—is the *identity* of particles. Even with the strictest standards of manufacture, no two ball bearings can ever be truly identical. Yet we have good reason to think that all electrons are truly identical, all red up-quarks are truly identical, and so on. The fact that electrons obey the Pauli exclusion principle (which says that no two of them ever exist together in exactly the same state of motion) can be understood if the electrons are identical but cannot be understood if electrons differ from one another in any way. If there were infinitely many layers of matter to uncover, we would expect electrons to be as complex as ball bearings, no two of them alike, and each one requiring a vast array of information for its exact description. This isn't the case. The simplicity of the fundamental particles and their identity give strong reason to believe that we may be close to reaching the ultimate "reality" of matter.

I turn now to the granularity of some *properties* of things.

## Charge and Spin

One of the quantized properties of matter that I have already introduced is electric charge. Some observed particles have no charge. All others carry a charge that is an integral multiple (positive or negative) of the charge $e$ of the proton. Spin is another such property. It is either zero or an integral multiple of the spin of the electron, which, in angular-momentum units, is $(\frac{1}{2})\hbar$. For the particles—including the composite particles—the multipliers of $e$ and $(\frac{1}{2})\hbar$ are typically 0, 1, or 2, but everyday objects may have charges vastly greater than $e$ and angular momenta vastly greater than $(\frac{1}{2})\hbar$. Lumpiness, or quantization, means that there is a finite difference between allowed values of the property. It does *not* mean that there is only a finite number of possible values. Just as the even numbers 2, 4, 6, and so on have a finite separation but go on without limit, charges and spins have a finite separation but an infinite number of possible values.

We have no understanding of the particular magnitude of the charge quantum $e$. It is relatively small, even by the standards of the particle world. It measures the strength of the interaction between charged particles and photons. This interaction (the electromagnetic interaction) is about a hundred times weaker than the quark-gluon interaction (understandably called the *strong interaction*).* On that basis we call the electron charge small, although both of these interactions are enormously strong relative to the weak interaction. But the point here is that the size of the lump of charge is simply a measured quantity. We don't know why it has the value it has.

Similarly, the magnitude of the quantity $\hbar$ (which sets the size of the lumps of spin) is a measured quantity without a theoretical underpinning. The theory of quantum mechanics, as developed in the 1920s, accounts for the *existence* of spin and angular-momentum quantization, but not for the magnitude of the quantum unit.

Bohr, in his 1913 paper, had postulated that angular momenta come in integral multiples of $\hbar$. Quantum theory later offered three rules for the lumpiness of spin.

1. Fermions (such as leptons and quarks) have half-odd-integral spin in units of $\hbar$ ($\frac{1}{2}$, $\frac{3}{2}$, $\frac{5}{2}$, and so on), while bosons (such as photons and gluons) have integral spin in this unit (0, 1, 2, and so on).

---

* One can't compare *exactly* the relative strengths of different interactions, because they have different mathematical forms. Roughly, it's as if you were attracted much more strongly to person A than to person B when you see them across the room but are attracted only a little more strongly to person A than to person B when you are close to them. The "relative strength of attraction" can be specified approximately, but not exactly.

2. Orbital angular momentum is always an integral multiple of $\hbar$ (0, 1, 2, and so on).

3. Angular momentum, either spin or orbital, can point in only certain directions, and the projections of angular momentum along any chosen axis differ from one to the next by exactly $\hbar$ (that is, one unit).

The third rule is especially interesting. When we say that spin "points" in a certain direction, we mean that the *axis* of spin points in that direction. We could say, for instance, that the spin of the Earth points toward the North Star, meaning that the Earth's axis points in that direction. The idea of "projection" is illustrated in Figure 16. The arrows represent angular momentum pointing in various directions. The lines drawn from the tips of the arrows perpendicularly to an axis depict projections of the angular momentum on that axis. That sounds a bit complicated, but the implications are easy to understand. The "quantization of orientation" (Rule 3) means that there are a limited

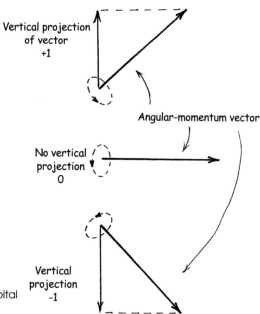

Vertical projection of vector +1

Angular-momentum vector

No vertical projection 0

Vertical projection -1

Figure 16. Projections of orbital angular momentum 1.

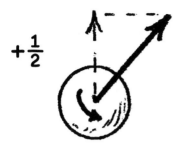

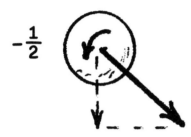

Figure 17. Projections of spin $\frac{1}{2}$.

number of possible orientations for any given angular momentum. For example, if a particular orbital angular momentum is 1, it can point "up," "down," or halfway between, to have three possible orientations. An electron or other spin-one-half particle has only two possible orientations, "up," with a projection of $\frac{1}{2}$, and "down," with a projection of $-\frac{1}{2}$ (according to Rule 3, the two projections must differ by 1).*

## Color Charge

Color charge, like electric charge, comes in quantum units: it is lumpy. As I discussed in Chapter 4, a quark can have any one of three "colors," conventionally called red, green, and blue. The colors available to anti-

---

* There is a mathematical formula that gives the number of possible orientations for any given angular momentum (in units of $\hbar$). Double the angular momentum and add 1. For spin one-half, this formula gives 2; for spin 1, it gives 3; and for spin 2, it gives 5.

quarks are called antired, antigreen, and antiblue. If two red quarks were to join together (in practice, they don't), the combination would have a color charge of two red units. Three red quarks in combination would have a color charge of three red units, and so on. This is the same as the rule for charges of the same sign. Two electrons have a total charge of $-2$ units, three have a total charge of $-3$ units, and so on. The "cancellation" of color is a bit more complicated. Whereas a combination of equal positive and negative charges has zero net charge, it takes a combination of three colors to create a colorless entity—either red, green, and blue, or antired, antigreen, and antiblue. As it turns out, the things we observe in nature are all colorless. This means there is no accumulation of color that makes itself felt in the large-scale world. By contrast, electric charge can, and does, accumulate in great quantity to produce the charged objects we see around us—whether it be a comb you have run through your hair or a thundercloud about to link to Earth with a bolt of lightning.

## Mass

In a way, the most obviously lumpy quantity is mass, for every particle has its own specific mass. Indeed, every composite entity—an atomic nucleus, for example, or a protein molecule—has its own definite (and therefore quantized) mass. But because of the contribution of energy to mass, the mass of a composite particle is not equal to the sum of the masses of its constituent particles. A single neutron, for example, has a mass that is considerably greater than the sum of the masses of the three quarks (plus any number of massless gluons) of which it is composed. Energy is part of the mix, and that energy contributes to the mass. Or consider a deuteron (the nucleus of heavy hydrogen), composed of a proton and a neutron. Its mass is a little *less* than the sum of the masses of a proton and a neutron. Energy again—but this time it is *binding energy*, a negative contributor to mass. In order to break up a deuteron into its constituent particles, energy must be added—just enough of it to offset the binding energy.

Charge and angular momentum are simpler. The charge of a neutron (zero) is the sum of the charges of the quarks that compose it. The deuteron's spin is the sum of the spins of the proton and neutron that make it up (added as vectors, taking account of direction as well as magnitude). And so on, for all combinations of particles. The fact that component masses do *not* add is a reminder that at the deepest level, a composite entity is not a simple combination of parts—it is a new entity entirely.

If you look at Table B.1 or Table B.2 in Appendix B, you will not see any apparent regularity in the masses of the fundamental particles. Scientists have not found any either (although approximate regularities are known for some composite particles). Why is the mass of the muon more than two hundred times the mass of the electron? Why is the mass of the top quark some fifty thousand times the mass of the up quark? No one knows. Quantized mass is just there, awaiting explanation.

## Energy

I turn finally to energy, that most ubiquitous of all physics concepts. The quantization of energy is where quantum theory began, and it played a role at every step thereafter. Planck and Einstein dealt with radiant energy. Bohr added material energy. Light radiated by the hydrogen atom had been known for many decades to produce a *line spectrum*, meaning that the atoms radiated only certain discrete frequencies (which appear as lines because the light is passed through a narrow slit before its frequencies are sorted). In prequantum days, there was nothing surprising about this. It was interpreted to mean that charge within the atom vibrated only at certain frequencies, just as air in an organ pipe or an oboe or a flute does, or as a piano string or violin string does. You might say that the atom radiated music, not noise.

By the time of Bohr's work, two developments made the atom's line spectrum seem more troubling, something crying out for a new explanation. First was the energy-frequency link of Planck and Einstein. If the atom was emitting only certain frequencies, it must be losing only certain specific amounts of energy. Second was the discovery by Ernest Rutherford and his colleagues, in 1911, that the atom is mostly empty space, consisting of a small central nucleus surrounded by electrons.

Classical theory was unable to cope with this situation. It predicted

Niels Bohr (1885–1962), 1922. Photo courtesy of AIP Emilio Segrè Visual Archives, W. F. Meggers Gallery of Nobel Laureates.

that an electron circling the nucleus in a hydrogen atom would spiral inward, radiating energy continuously at ever-higher frequencies and falling into the nucleus in about $10^{-8}$ second. The twenty-six-year-old Bohr, on leave from his native Denmark as a guest researcher at Rutherford's lab in Manchester, England, could see that radically new ideas were needed. The electron in a hydrogen atom, he reasoned, couldn't lose its energy gradually and continuously. It must exist for some time in a *stationary state*, then make a *quantum jump* to a lower energy stationary state, until finally it reached a *ground state*, from which it could radiate no further. Three revolutionary ideas, all tied together. When to these he added a fourth revolutionary idea, that angular momentum is quantized in units of $\hbar$, he could account quantitatively for the observed frequencies in the line spectrum of hydrogen.

Bohr introduced and used yet another idea, the *correspondence prin-*

*ciple*, which says that when quantum granularity (lumpiness) becomes relatively small, quantum results should be close to classical results. Let me explain what the unscientific-sounding phrases "relatively small" and "close to" mean. As applied to the hydrogen atom, "relatively small" means that the *fractional* change between adjacent quantized quantities is small. If the energies of two adjacent states or the angular momentum of two adjacent states differ by, say, 1 percent, we would say that the granularity is relatively small. If the difference between the energies of adjacent states is one-hundredth of 1 percent, the granularity is, relatively, even smaller. It turns out that for the lowest-energy states of the electron in the hydrogen atom, the quantum steps are relatively large. Then classical reasoning fails completely. But as the electron circles farther and farther from the nucleus in what are called "excited" states, the *relative* change from one to the next gets smaller and smaller. Then classical reasoning begins to have some validity. When we say that for these states quantum results are "close to" classical results, we really mean that as the quantum granularity gets smaller and smaller (percentagewise), the difference between quantum and classical results also gets smaller and smaller. Quantum results then "correspond" to classical results.

The correspondence principle is imprecise. We might even call it fuzzy. But it is a powerful principle, for it puts constraints on quantum theory. It requires that when quantum theory is applied to situations where classical theory has already been found to be valid, the quantum results must duplicate (closely) the classical results. It's a bit like the transition from a ten-lane superhighway to a set of city streets. The traffic engineer must arrange for the traffic flow to change gradually and smoothly from one regime to the other. The engineer's "correspondence principle" is the requirement that as the number of lanes get smaller and smaller, the freeway rules of the road must approach the city rules of the road.

Bohr submitted a draft of his paper to his mentor Rutherford (then forty-one), telling Rutherford that he was reluctant to publish because he could explain only the spectrum of hydrogen, not the spectrum of any heavier element. Rutherford wisely counseled him to go ahead and publish, reportedly saying, "If you can explain hydrogen, people will be-

Ernest Rutherford (1871–1937), ca. 1906. Photo courtesy of AIP Emilio Segrè Visual Archives, gift of Otto Hahn and Lawrence Badash.

lieve the rest." But one thing about Bohr's paper did bother Rutherford.* "It seems to me," he wrote, "you would have to assume that the electron knows beforehand where it is going to stop." Rutherford's puzzlement gets right at a quantum mystery that is the subject of the next chapter.

An electron in a "stationary state" is not literally stationary, not motionless. It moves at high speed throughout some region of space, but its energy and angular momentum have fixed (stationary) values. An electron cascading from one energy state to a lower-energy state, then to a still lower-energy state, and so on, is like a person walking down a flight of stairs, pausing at each step. The person loses some potential energy in going from one step to the next lower one, waits, then loses some more potential energy, until reaching the floor at the bottom of the steps. The electron also reaches a floor. There is a lowest-energy state, the ground state. And this ground state has *zero-point energy*, energy that can't be

---

* Rutherford, a big, gruff New Zealander who loved to sing lustily in his laboratory, had no patience with mathematical theory that didn't link to something he could picture. Bohr's stationary states and quantum jumps, revolutionary though they were, met Rutherford's test.

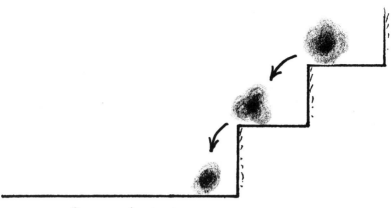

## Ground state

Figure 18. Electron cascading to a lower-energy state.

extracted. The electron is still zipping around vigorously with lots of kinetic energy, but the rules of quantum mechanics prevent it from losing any of that energy by falling to a lower-energy state, for there *is* no lower-energy state. It's as if when you reach the floor at the bottom of a flight of stairs, there is nowhere lower for you to go. You are prohibited (let's say) from digging into the Earth. You are in your ground state. Even though there is, in principle, lots of additional energy that you could release by, for instance, moving downward in a mine shaft toward the Earth's center, you are prevented from doing this. You remain perched at ground level.

This concept of zero-point energy appears also when we consider the cooling of materials to absolute zero. Absolute zero is not truly reachable, but physicists have achieved temperatures within less than a millionth of a degree of it. We may define absolute zero as the temperature at which no further heat (energy) can be extracted from a material. The material as a whole is in its ground state, just as a single atom can be. Its energy there (which might be considerable although unusable) is its zero-point energy.

It turns out that quantum theory leads to quantized energy only for systems that are confined in space—systems like atoms and molecules and atomic nuclei. As a general rule, the more confined the system, the greater the energy separation of the states of motion.* Thus, energy states in tiny nuclei are widely separated, typically by hundreds of thousands or millions of electron volts (eV), whereas states in much larger atoms are typically separated by only a few eV. Molecular energy spacings can be smaller still, because molecules are larger than atoms.

A corollary of the above rule is that if there is no confinement, there is no separation of permitted energy states. An electron flying freely through space, being unconfined, can have any energy at all—or, to put it differently, the energy separation between its permitted states of motion is zero. In a piece of metal, a conduction electron (one responsible for carrying electric current) is loosely confined, for it may be free to move over a distance of only a few centimeters. But that is such an enormous distance relative to atomic or nuclear dimensions that the conduction electron is, for all practical purposes, unconfined. The separation between its adjacent energy states is too tiny to matter.

Now to a subtle point that relates to fundamental particles. Are two hydrogen atoms, one in its ground state and one in an excited state, two versions of the same entity or are they two different entities? We usually treat them as two versions of the same entity because they have so much in common. But from the deepest perspective, they are different entities. An excited hydrogen atom that emits a photon and turns into a hydrogen atom in its ground state is, in principle, behaving no differently from a muon that emits a neutrino and an antineutrino and "becomes" an electron. In the latter process—the particle process—we normally say that the muon has been annihilated and the neutrino, antineutrino, and electron have been created. With equal justification, we could say that when an excited hydrogen atom decays, it is annihilated and a photon and a hydrogen atom in its ground state have been created. So, in the end, whether we are talking about atoms or molecules or nuclei or particles, processes of change are processes in which one or more entities are annihilated and one or more other entities are created.

The reasoning above leads us to ask a reverse question. If different

---

* The explanation for this rule is to be found in the wave nature of matter, which I explore in Chapter 9.

energy states of an atom can be regarded as distinctly different entities, can distinctly different particles be regarded as different energy states of some underlying "thing"? For some composite particles, made of the same set of quarks, the answer is certainly yes. We know that the same quarks can combine in different ways—for instance, with differently aligned spins—to make different particles. The masses of these particles reflect their total energies, and physicists have found approximate relationships among the masses of such composite particles, analogous to the relationships among the energies of the excited states of the hydrogen atom. For what we call the fundamental particles, on the other hand (leptons, quarks, and force carriers), we do not know the answer to the question. So far, these particles seem to be quite distinct. It remains to be learned whether, through string theory or in some other way, even the fundamental particles are "just" different energy states of some deeper underlying thing.

## chapter 6

# Quantum Jumps

Rutherford's question to Bohr was a good one: How does an electron in an excited state know to which lower-energy state it should jump?* And I can add another question: How does the electron know *when* to jump? More than a dozen years passed before these questions were answered. In 1926, Max Born†—part of the group in Göttingen, Germany, that was instrumental in creating the new quantum mechanics—had the insight that led to the answer. The fundamental laws of the subatomic world, he said, are laws of probability, not laws of certainty. Quantum mechanics does not, and cannot, predict when a particular electron in a particular atom will make a quantum jump or to what state it will jump. In other words, the electron *doesn't* know when or where to jump. What quantum mechanics predicts is the *probability*

---

* Of course, the electron doesn't know anything. This terminology is just the physicist's way of talking about the laws governing the electron.

† Born left Nazi Germany in 1933, going first to England, then, in 1936, to Edinburgh, Scotland, where he held a professorship until his retirement in 1953. His granddaughter Olivia Newton-John chose singing and acting instead of physics.

Max Born (1882–1970). Photo
courtesy of AIP Emilio Segrè
Visual Archives.

that the electron will jump. That probability can be calculated quite
precisely. So we know *exactly* the probability that an electron in an excited state **A** will make a quantum jump to some lower-energy state **B**.
For any particular atom, we have no way of knowing just when that
quantum jump will occur or even if it will occur at all, since the electron
might instead jump to some other state **C**.

How, you might ask, can probability be exact, when the essence of
probability is inexactness, or uncertainty? To answer this question, let
me turn to the everyday world. The probability that a perfectly balanced coin, when flipped, will come up heads is $\frac{1}{2}$, or 0.5. This is an *exact* probability (given a perfectly balanced coin). The chance of heads is
not 0.493 or 0.501, it is precisely 0.500. Yet the result of any given coin

toss is completely uncertain. So probability can be exact when outcomes are uncertain.

Does the coin itself know whether it is going to land heads or tails, even if you, the coin flipper, don't know? You probably think that the coin, like the electron, doesn't know. After all, isn't the flipped coin a perfect demonstration of probability at work? Yes (in practice) and no (in principle). The football referee who flips a coin doesn't know whether it will land heads or tails, but the *coin* knows—meaning that a scientist who knew everything there was to know about the coin flip (such as the height of launch; the coin's mass, initial speed, direction, and rate of rotation; the wind speed; and the air resistance) could, in principle, calculate which face will be up when the coin lands. What we have with the flipped coin is a *probability of ignorance*. The reason we can, in practice, give only the probability of heads or tails (50 percent each) and not predict the outcome of any particular coin flip is that we don't know enough. We are ignorant of the details that would be necessary in order to calculate the outcome. The quantum probability governing the electron is different. It is a *fundamental probability*. We can know everything there is to know about an electron in an excited state and still not be able to predict when or where it will jump.

Classical physics is unambiguous and exact—in the sense that every outcome, under all conditions, can, *in principle*, be calculated precisely. It's just a matter of knowing enough about what are called *initial conditions*. We know enough about present conditions in the solar system to predict with confidence exactly where the planet Mars will be on January 1, 2050. Even next week's weather could (in principle) be calculated precisely if we knew everything there is to know about the air and the Earth beneath it today.* Quantum mechanics is also unambiguous and exact, but in a different sense: in the sense that *probabilities* can be cal-

---

* Unfortunately, this is beyond all hope of realization—not just because of the difficulty in getting enough information about today's weather, but because next week's weather is incredibly sensitive to the tiniest uncertainties in today's weather. The phenomenon called *chaos* magnifies today's tiny variations into next week's giant effects.

culated precisely. It is ambiguous and inexact (or, better, uncertain) in that a physicist can calculate only the *chance* that something will happen, never what will in fact happen. Exactly when an electron chooses to make its quantum jump remains beyond calculation.

What measurements can the physicist make to find out if the calculated probabilities are really correct? To measure the probability, say, that an electron in a hydrogen atom will jump from state **A** to state **B,** measurements on a single atom, or even a few atoms, will never do. One must study a great many atoms, all initially in the state **A,** and observe the *average* behavior of those that jump to state **B.** Each one of them will decay (that is, make its quantum jump) after a certain time. The measured times may cover a wide range and may all be different. If the experimenter is patient and measures the decay times for a million atoms, he or she can find the *average* of those times, which we call the *mean lifetime*. This measured mean lifetime can be compared with a theoretical mean lifetime to check on the validity of the theory.* After a million measurements, there should be close agreement (if the theory is correct), even though any individual lifetime may be very different from the average.

Similar reasoning applies to coin flipping. To check whether the probability of heads is really 50 percent, you would also need a great many measurements. If you flipped a perfectly balanced coin ten times, you might find that it came up heads only three times. That wouldn't tell you that the chance of heads is 30 percent. It would tell you only that you hadn't flipped the coin enough times. If you flipped it a thousand times, you would be surprised indeed if it came up heads only three hundred times. You would expect to see heads "almost" five hundred times. And if you flipped the coin a million times and found that heads

---

* Mathematically, the mean lifetime is the inverse of the decay probability. Suppose, for example, that the calculated decay probability is 20 percent per nanosecond (meaning that in any one nanosecond, there is a 20-percent chance that the decay will occur). This probability can be written 0.2/nanosecond. Its inverse is 5 nanoseconds. That is the predicted mean lifetime.

showed up 499,655 times, you would be prepared to believe that the probability of heads is 50 percent.

As I discussed in the previous chapter, the emission of a photon by an atom as it jumps from a higher- to a lower-energy state does not differ in principle from the decay of an unstable particle. A quantum jump, like a particle decay, is really a mini-explosion, in which what was there "before" vanishes, to be replaced by what is there "after." Certain quantities like energy are conserved (the same after as before), but little else is preserved. Just as we can represent pion and muon decay by

$$\pi^+ \to \mu^+ + \nu_\mu \text{ and } \mu^+ \to e^+ + \bar{\nu}_\mu + \nu_e,$$

we can represent the atomic transition from state **A** to state **B** by

$$\mathbf{A} \to \mathbf{B} + \gamma$$

(where $\gamma$ stands for a photon).

So the decay of an unstable particle, like the decay of an excited atom, is governed by probability. The time when any particular pion decays is completely uncertain, but a large number of pions have a mean lifetime (which, as shown in Table B.3 in Appendix B, is $2.6 \times 10^{-8}$ s).

Probability governs both *when* something happens and (if there is more than one option) *what* happens. In an atom, the available energy states (the so-called *stationary states*) can be represented by a ladder, as shown in Figure 19. The lowest-energy state is labeled "Ground." The higher-energy states are represented by the ladder's rungs. Suppose that an atom finds itself on the third rung. Again personifying the electron, we can say it has two options to consider: when it will decay and to which lower-energy state it will decay. There is a certain probability of its decay to each lower state. But to which state it actually decays and how long it waits before doing so are completely unpredictable.

A pion (to pick a particle example) also has options. It turns out that the decay of the pion into a muon and a neutrino, as indicated symbolically above, is overwhelmingly likely. It occurs 99.988 percent of

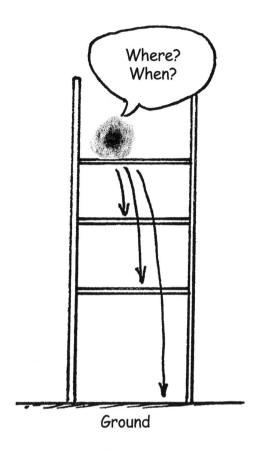

Figure 19. The energy "ladder."

the time. But occasionally a pion decays into three particles, including a photon,

$$\pi^+ \to \mu^+ + \nu_\mu + \gamma,$$

or into an electron (in this example a positron) and a neutrino,

$$\pi^+ \to e^+ + \nu_e,$$

The relative probabilities of these different decay modes are called *branching ratios*.

The idea that the fundamental processes of nature are governed by laws of probability, not laws of certainty, should have hit the world of science like a bombshell. After all, that single idea toppled the solid edifice of classical physics, constructed so laboriously over three centuries. But as it turned out, it was erosion rather than explosion that brought down the classical structure. Only after the mathematical theory of quantum mechanics was developed in the mid-1920s could Born state clearly its probabilistic interpretation.*

As early as 1899, Ernest Rutherford and others studying the newly discovered phenomenon of radioactivity noticed that the decay of radioactive atoms seemed to follow a law of probability. Exactly as with the example of excited atoms or pions, some lived a short time, some a longer time. Only the average time for any group was constant. Moreover, a single radioactive atom might have a choice of ways to end its life—for example, by emitting an alpha particle or by emitting a beta particle. For any given atom, the choice was unpredictable. Only the observation of many decay events allowed for the measurement of the relative probabilities (the branching ratio). Yet Rutherford and his fellow physicists didn't shout from the housetops that the fundamental laws of nature must be laws of probability. Why not?

The answer is very simple. They did not realize that they were dealing with *fundamental* laws. Probability in science was nothing new. What was new, but not yet recognized, was that for the first time probability was appearing in simple elementary phenomena in nature.

Rutherford assumed, no doubt, that he was dealing with a probability of ignorance. The interior of the atom was, as far as he knew, a complicated place, so that the apparent randomness of the decay process might have been due to unknown differences in the internal states of different atoms. (He had no inkling yet that atoms contained nuclei which were the seat of radioactivity.) During the first quarter of the twentieth century, there were other hints that probability might be fundamental, but so radical a notion was not likely to be accepted until experiment and theory converged to force it on science. Rutherford discovered (with Frederick Soddy, in 1902) that radioactivity represents a

---

* In 1924, two years before Born's work, the Dane Niels Bohr, the Dutchman Hendrik Kramers, and the American John Slater, working together, had suggested that probability might play a fundamental role in quantum processes. But they didn't yet have the completed theory that could give substance to the suggestion.

sudden catastrophic change in the atom, not gradual change. This, in itself, made radioactive transmutation* appear to be a rather fundamental event. Einstein's discovery in 1905 that light can be absorbed only in discrete bundles (photons) and Bohr's 1913 theory of quantum jumps in the hydrogen atom also contained at least flags of warning that perhaps probability is working at a fundamental level. But the world of physics wasn't yet ready to heed the warnings.

The probability that I have been describing so far manifests itself in the randomness of subatomic events. This randomness can show itself in various ways: in the *lifetime* of an excited atom or unstable particle, in the *branching ratio* to different possible outcomes, and also in what is called *scattering*. If one particle flies close to another, it can be deflected, or "scattered." Quantum mechanics permits only the calculation of the probability of a certain deflection, never the certainty of a deflection. A great deal of what we know about particle interactions has been learned through scattering experiments.

Not everyone can readily lay hands on a Geiger counter and a weak radioactive source, although they are common in high-school and college labs. At the heart of a Geiger counter is a metal tube containing a dilute gas, with a metal wire running along the axis of the tube. Between the tube and its central wire a high voltage is applied (hundreds of volts), almost but not quite sufficient to make a spark jump. When a high-energy particle flies through the tube, it ionizes molecules of the gas—that is, ejects electrons from the molecules—creating an electrified gas that enables the spark to jump. The brief pulse of current between the tube and the wire is amplified by an external circuit that can cause an audible click and/or cause a counter to register one more count. The circuitry also quenches the spark in a tiny fraction of a second, after which the tube is ready for the next particle. (Rutherford's postdoctoral student Hans Geiger invented the first primitive form of this detecting device around 1908 and perfected it later.)

Counting radioactive-decay events is a wonderful way to come into

---

* Transmutation, the dream of alchemists in the Middle Ages, is the change of one element into another. This happens in any radioactive transformation in which the charge of the nucleus changes.

direct personal contact with fundamental probability. If you hold the Geiger counter at such a distance from the radioactive source that you hear a click each time a high-energy particle penetrates the counter, you will notice at once that the clicks are not coming in regular sequence like the ticks of a clock. They seem to be random. Indeed, a mathematical analysis would show that they are exactly random. The time at which a given click occurs is completely unrelated to the time elapsed since the last click or the time at which any other click occurred. You, the listener, a truly gigantic creature by atomic standards, are hearing individual messages from the subatomic world. Every audible click means that somewhere, among the countless billions of atoms in the radioactive sample, one nucleus has suddenly decided to eject a particle at high speed, transmuting itself into a different nucleus.* Very literally, a nuclear explosion has occurred, and, in the private world of the nucleus, the time of the explosion has been governed exclusively by a law of probability. A neighboring nucleus may have long since exploded, or it may continue to live a long time.

Probability manifests itself in another way, which is not so obvious to the eye or ear but is equally convincing to someone with a little mathematical training. This is through the law of exponential decay. Rutherford discovered the role of probability in radioactivity in this way, for in 1899 he had no means of observing single transmutation events. Rutherford noticed that when the intensity of the radioactivity was graphed as a function of time, a curve like the one in Figure 20 resulted. This is called an *exponential curve*. The most marked characteristic of such a curve is that it falls vertically from any value whatever to half that value in a fixed horizontal distance. This meant in Rutherford's experiments that a definite fixed time was needed for the radioactivity to diminish in intensity by half, regardless of the initial intensity. This fixed time is called the *half-life* of the material.

What Rutherford knew and what I must ask the reader to accept is that the exponential curve results from the action of a law of probability on the individual radioactive-decay events. For each single nucleus, the

---

* It is not strictly true that all of the clicks you hear are produced by radioactive decay events. Some are caused by cosmic-ray particles, mostly muons, that happen to be streaking down from above.

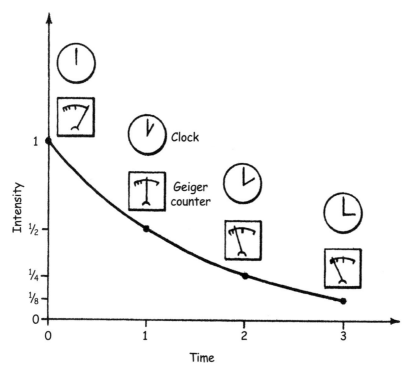

Figure 20. Exponential decay of radioactive sample.

half-life represents a midway point in probability. The chance that the nucleus will decay in less time is one-half; the chance that it will decay after a longer time is one-half. When this probabilistic law acts separately on a large collection of identical nuclei, the total rate of radioactive decay falls smoothly downward along an exponential curve. The same is true for particles. Each of the mean lives* shown in Tables B.1 and B.3 was measured by studying the exponential decay curve for the

---

* In general, a mean life, or average life, is not the same as a half-life. In 2000, for example, the *average* life expectancy (mean life) in America was 77 years (for men and women combined), and the *half*-life was 80 years. This means that not until reaching age 80 has an American outlived half of his or her contemporaries. In the particle world, on the other hand, where the probability of "dying" is the same at all ages, the half-life is considerably less than the mean life (0.694 is the exact factor connecting them). A neutron, with a mean life of fifteen minutes, has outlived half of its contemporary neutrons after about ten minutes.

particular particles. (The times themselves, though not measured directly, can be inferred by measuring the speeds of the particles and the distances traveled.)

The span of known half-lives from the shortest to the longest is unimaginably great—from less than $10^{-22}$ s to more than $10^{10}$ years. Regardless of half-life, the decay of every unstable particle or nucleus follows the same inexorable exponential curve.

You can picture how, with good timepieces, a scientist can measure half-lives ranging from seconds to years. You can probably guess that with modern timing circuits, it is possible to measure time intervals as short as millionths or billionths of a second. By using distance as a surrogate for time, physicists can determine lifetimes of a trillionth of a second ($10^{-12}$ s) or less. In $10^{-12}$ s, a particle moving near the speed of light covers about a third of a millimeter (or more, if the relativistic dilation of time is taken into account). But how in the world can scientists measure half-lives of $10^{-22}$ s, during which time a particle can move only a small fraction of the distance across an atom? Or how can they measure a half-life that is greater than the time elapsed since the Big Bang? The answer to the second question (how to measure very long lives) is the more straightforward. It depends on the working of probability. Suppose that you have a radioactive atom whose mean life is a billion years. This means that it has a one-in-a-billion chance of decaying in a single year. If you have a billion such atoms, one of them will, on average, decay each year. If you have 365 billion of them, one will, on average, decay each day. But 365 billion is a very small number of atoms. In practice, a radioactive sample can contain so many atoms that many will decay each second, even if the mean life is a billion years. So it's an easy matter to measure very long half-lives, thanks to the fact that the atoms are being pressed by probability every moment. A small percentage of them will decay well before they are old.

Measuring extremely short half-lives takes advantage of a wholly different feature of quantum mechanics: the *uncertainty principle*. The shorter the lifetime of a particle, the greater the uncertainty of its energy. The energy uncertainty gives a measurable "spread" of energy. From that measured spread, the lifetime can be deduced.

Nuclear waste may seem to be a topic far removed from quantum principles, but in fact it is a public-health concern that is directly related to quantum jumps and probability. In what is called *spent fuel*, there are myriad radioactive isotopes with half-lives ranging from seconds to thousands of years. Each of them follows its own exponential decay curve. Overall, the radioactivity of the mixture diminishes very rapidly at first, more slowly later. There is no specific time after which the once unsafe material becomes safe. Rather, there is a gradual change from more hazardous to less hazardous. Since the seat of radioactivity is the atomic nucleus, no chemical or physical treatment of any kind can affect it. For better or worse, it is there, and we must cope with it. Some people have suggested that radioactive waste be loaded onto rockets and blasted into space to fall into the Sun. Since this would be enormously expensive and would pose great danger on launch, it hardly seems practical. Another futuristic possibility—which may one day become practical—is to "cook" the hazardous material in a *fusion furnace*, an incinerator so hot that it uses nuclear rather than chemical reactions to break up the material into innocuous forms. In the nearer term, the radioactive material will have to be stored on Earth in such a way that it is segregated from the human environment for centuries to come.

A particularly fascinating kind of quantum jump is a jump from one side to the other of an "impenetrable" barrier, a phenomenon called *tunneling*. Like other kinds of quantum jumps, tunneling is governed by a law of probability. If a particle is held on one side of a wall which according to classical physics is totally impenetrable, there is a certain small chance that it will pop through, emerging on the other side. As I mentioned in Chapter 3, the alpha decay of nuclei can be explained as a tunneling phenomenon. The alpha particle is held within the nucleus by a wall of electric force that, in classical theory, cannot be breached. Yet, with a certain small probability, the alpha particle can suddenly make an appearance outside the nucleus and fly away, to be recorded in someone's particle detector.

Tunneling usually occurs with exceedingly small probability. Within a nucleus, an alpha particle may be "knocking on the door" $10^{20}$ times per second to get out, yet may make it through only after millions of years. In our human-sized world, the probabilities are even smaller. A warden need have no fear that a prisoner leaning against the prison wall will suddenly be free on the other side of the wall. A student in a dull lecture has no hope of popping up in the lounge outside the lecture hall. Tunneling helps particles escape from the confines of their nuclear prisons, but will never help you escape from where you are.

Yet in recent decades scientists have learned how to engineer tunneling. If two materials are maintained at a small difference of electrical potential (often less than one volt) and are brought very close together, electrons in one of the materials may "tunnel" through the gap between them even though, again, classical physics says there is no way for the electrons to get from one material to the other. A beautiful application of this technique is the scanning tunneling microscope (STM), perfected by Gerd Binnig and Heinrich Rohrer in an IBM lab in Zurich— an achievement for which they shared a Nobel Prize in 1986.

The STM is not a microscope in any usual sense, but it is nevertheless aptly named, for it can produce images of a solid surface that reveal the locations of single atoms. The surface being studied is brought close to a fine metal tip that can be moved back and forth above the surface (thus the term "scanning") and can also be moved up and down—that is, toward and away from the surface—with incredibly precise control. The tip is brought to within a nanometer ($10^{-9}$ m, less than ten atomic diameters) of the surface. At that distance, electrons tunnel from the surface to the tip and can be recorded as a weak electric current. Since the surface, with its lumpy atoms, is not exactly flat, the distance from surface to tip will vary as the tip is moved laterally above the surface. As the distance increases slightly, the tunneling current decreases. As the distance decreases slightly, the tunneling current increases. To make this device function as a microscope, feedback circuits move the horizontally scanning tip ever so gently toward and away from the surface in such a way as to maintain a constant tunneling current—which means a constant distance from tip to surface. So the recorded up-and-down motion of the tip can be translated into a map of the hills and valleys of the surface. The map is accurate to less than the diameter of a single atom, about a tenth of a nanometer ($10^{-10}$ m).

Another engineered application of tunneling is the *tunnel diode*. A diode is a device through which electric current flows readily in only one direction. It is rather like a turnstile that lets you leave a subway

Heinrich Rohrer (b. 1933), left, and Gerd Binnig (b. 1947) with their first scanning tunneling microscope (STM). Photo courtesy of IBM Zurich Research Laboratory.

platform or enter a zoo but doesn't let you go through it in the opposite direction. Diodes are common in electronic circuits and are usually made of two semiconductors in contact. With the right choice of materials and with a certain chosen electric potential between the two pieces, electric current can be stopped completely, according to classical calculations. But if the boundary layer between the two materials is thin enough, some electrons can tunnel through it. Tunnel diodes are in common use because they happen to have an interesting property called *negative resistance*. This means that within a certain range of voltage, more voltage produces less current, contrary to the usual state of affairs.

Whenever a quantum jump occurs, it is "downhill"—that is, from a state of higher mass to a state of lower mass. This downhill rule is evi-

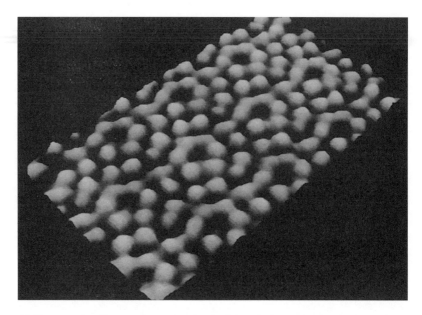

STM image of a silicon surface, showing individual atoms. Photo courtesy of IBM Zurich Research Laboratory.

dent for particle decay—for instance, in the decay of a lambda particle into a proton and a negative pion, indicated by

$$\Lambda^\circ \to p^+ + \pi^-.$$

From Table B.3, we get

mass before $= 1{,}116$ MeV;
mass after $= 938.3$ MeV $+ 139.6$ MeV $= 1{,}077.9$ MeV.

The mass decreases by about 38 MeV, or 3 percent. For the decay of a neutral pion into two photons, the mass decrease is 100 percent, since the products of the decay are massless. The downhill rule applies even to the quantum jump of a hydrogen atom from a higher-energy to a lower-energy state, although then the percentage mass change is tiny. When the electron in a hydrogen atom jumps from its first excited state to its ground state, the atom loses about 10 eV in mass. This is only one part in a hundred million (or a millionth of one percent) of the atom's mass. The change is so small that it has never been directly measured, and, prior to the development of relativity and quantum mechanics, had not even been guessed at. But scientists have measured so many other mass changes resulting from quantum jumps, including those in radioactive decay, that there is no doubt the downhill rule applies even to atoms.

Why the downhill rule? Picture yourself on skis on a mountainside. The only direction you can go spontaneously (that is, without adding any energy) is down. You obey a "downhill rule." Energy conservation prevents you from wafting effortlessly up the slope, just as it prevents a radioactive particle from transforming into products that have more mass than it does. For a radioactive particle, at rest and alone, the *total energy*, both before and after its decay, is equal to its mass energy. Some of this energy goes into the masses of the product particles, and some goes into their kinetic energy. This means that the product particles can have no more mass than the parent particle—and, in practice, always have less.

You might think that if the radioactive particle is moving, some of its kinetic energy could be transformed into mass, resulting in possible "uphill" decay. Relativity helps to explain why this won't work. Climb into a frame of reference moving with the radioactive particle. In that frame of reference, the particle is at rest. And in that frame, the downhill rule applies. If the masses of the product particles add up to less than the mass of the parent particle in the moving frame of reference, they add up to less in all frames of reference, for particle mass doesn't change from one frame of reference to another.*

Let me return for a moment to your imagined position on skis on a mountainside. In fact, you *can* go uphill. You can board a ski lift, or you can climb, transforming some of your stored chemical energy into potential energy. A particle, too, can go "uphill" if energy is added to it in some way. This is exactly what happens in accelerators. One particle is struck by another, resulting in the possible transformation of some ki-

---

* In this discussion, mass means rest mass, which is an invariant quantity—that is, a quantity independent of the frame of reference.

netic energy into mass energy. So it is only for spontaneous decay that the downhill rule applies.

I have focused in this chapter on quantum jumps and the laws of probability that govern them. Yet not everything in the subatomic realm is uncertain and probabilistic. Many of the properties of stable systems—the spin of an electron and the mass of a proton, to give two examples—are well defined. Nevertheless, much of what happens to atoms, nuclei, and particles *is* subject to probability. This leads us to ask: If most of what happens in the small-scale world is governed by laws of probability, why isn't the same true in the large-scale world? The large-scale world is, after all, built of numerous pieces of the small-scale world and must be subject to the same laws. (As I remarked earlier, we often encounter probability in the large-scale world, but it always arises from lack of full information—it is a probability of ignorance, not a fundamental probability.) There are two reasons fundamental probability usually hides from our view in the everyday world. One reason is that when enough individual probabilistic events are amassed, the result can be smooth, predictable change. This is true in radioactivity, where the random decays of individual nuclei combine to create a smooth, exponential change. The other reason is that fundamental probabilities, when extrapolated to the large-scale world, often turn out to be nearly zero or nearly one. The chance that quantum-mechanical tunneling will propel you through a brick wall is effectively zero (but not *exactly* zero). The chance that a thrown baseball will follow a smooth path, not zig or zag unpredictably, is effectively one—that is, 100 percent (but not *exactly* one).

Finally, this deeper question: Are the quantum laws of probability *really* fundamental, or do they represent probability of ignorance masquerading as fundamental? Ultimately, no one knows, but the idea that quantum probability is fundamental has been around for more than seventy-five years, and it is holding up. One argument for it is simple and direct: it works. Another argument, less direct, has to do with an idea that I discussed in the preceding chapter—the idea that electrons are simpler objects than ball bearings, that there are no layers of deeper reality yet to uncover. If the probability that governs the time when an

"God is subtle, but not malicious" ("Raffiniert ist der Herr Gott / Aber Boshaft ist Er nicht"). This quotation from Einstein is carved above a fireplace in Jones Hall (formerly Fine Hall) at Princeton University. Photo courtesy of Denise Applewhite.

electron makes a quantum jump were a probability of ignorance, this would mean that the electron has many properties other than those we know, hidden properties that determine the time of the jump even though we scientists don't know enough to predict that time. It's like the steel ball in a roulette wheel. The reason we can't predict the slot in which the ball is going to land is that the ball and the wheel have many detailed properties we don't know, properties that determine the ball's final resting place and that would permit us to predict the resting place if we knew them all and laboriously calculated the motion. We have no evidence for such numerous unknown properties of the electron, and indeed good evidence that they don't exist. Therefore, we say that quantum probability is fundamental.

Nevertheless, many scientists still feel uneasy about quantum probability. It makes quantum mechanics weird. It goes against common sense. It conflicts with some philosophies. Albert Einstein, generally regarded as the greatest physicist of the twentieth century—and, ironically, one of the architects of quantum theory—never liked quantum probability. He often remarked that he did not believe God played dice, and in 1953 he wrote, "In my opinion it is deeply unsatisfying to base physics on such a theoretical outlook, since relinquishing the possibility of an objective description . . . cannot but cause one's picture of the physical world to dissolve into a fog."* Another famous Einstein quote is carved in stone in the Princeton University building where Einstein once had his office. It says, "God is subtle but not malicious." Einstein accepted God's subtlety: the laws of nature are not immediately evident, he was saying; it takes great effort to figure them out. But surely no grand designer of the universe would be so malicious as to build unpredictability into the basic laws.†

If a *reason* for quantum unpredictability is one day found, the fog may lift.

---

* A. Einstein, "Elementary Reflections on the Interpretation of the Foundations of Quantum Mechanics" [in German], in *Scientific Papers Presented to Max Born* (New York: Hafner, 1953), p. 40.

† Whether Einstein believed in God in any conventional sense is questionable, though he often alluded to the "Lord" or the "Old One."

# chapter 7
# Social and Antisocial Particles

In Gilbert and Sullivan's *Iolanthe*, Private Willis muses (in song) while standing guard at night,

> *I often think it's comical*
> *How Nature always does contrive*
> *That every boy and every gal*
> *That's born into the world alive*
> *Is either a little Liberal*
> *Or else a little Conservative!*

Nature has also contrived that particles have little choice:

> *It often seems quite strange to me*
> *(Please stop me if I carry on)*
> *That every physics entity—*
> *Atom, quark, or huge geon—*
> *Must either social boson be,*
> *or antisocial fermion.* *

---

* Thanks to Adam Ford for poetic inspiration.

Every particle—and indeed every entity built of particles, such as an atom or molecule—is either a boson or a fermion.* Earlier I introduced you to some particles of each type. Quarks and leptons (including the most ubiquitous lepton, the electron) are fermions. Photons and gluons and other force carriers are bosons. And there are many composite particles of both types. Protons and neutrons, for example, are fermions. Pions and kaons are bosons. Do you see a rule here? Entities made of an odd number of quarks are fermions. Entities made of an even number of quarks (or a quark-antiquark pair) are bosons. The more general rule is this: Something built of an odd number of fermions is itself a fermion. Something built of an even number of fermions or *any* number of bosons† is a boson. This may sound confusing, but here's a simple way to think about it. Imagine a negative sign attached to every fermion and a positive sign attached to every boson. Apply a negative sign an odd number of times and you get a negative. Apply a negative sign an even number of times and you get a positive. Apply a positive sign any number of times and you get a positive.

Here's a quiz: Is an atom of sodium 23, whose nucleus contains 11 protons and 12 neutrons, a boson or a fermion? You might think it is a fermion because its nucleus contains 23 fermions—an odd number. Or, if you count quarks—three for each proton and three for each neutron—the nucleus contains 69 fermions, still an odd number. But wait. Circling the nucleus are 11 electrons. So quarks plus electrons add up to 80, an even number. (Or protons, neutrons, and electrons add up to 34, also an even number.) The atom of sodium 23 is a boson.

Bosons and fermions possess many properties that don't distinguish one from the other. For instance, they can be fundamental or composite. They can be charged (positive or negative) or neutral. They can in-

---

* The geon cited in the poem is a hypothetical entity made of a vast number of photons bundled so closely together that they circle around a common point under the influence of their own powerful gravity. Yes, photons attract one another. All concentrated energy, be it massless or massive, exerts and feels gravitational force.

† No composite structures made of *fundamental* bosons are known, although a pair of "entangled" photons (see Chapter 10) can be considered a single entity. An example of a composite made of composites would be a molecule whose atoms are all bosons. Then the molecule as a whole is a boson.

teract strongly or weakly. They can have a wide range of masses, including (at least for the photon) zero mass. But when it comes to spin, they are different. Bosons have integral spin (0, 1, 2, and so on); fermions have half-odd-integral spin ($\frac{1}{2}$, $\frac{3}{2}$, $\frac{5}{2}$, and so on).

Their greatest difference, however, is how they behave when two or more of them are together. Fermions are "antisocial." They obey the *exclusion principle*, which states that no two identical fermions (for instance, no two electrons) can occupy the same state of motion at the same time. Bosons are "social." Not only can two identical bosons occupy the same state of motion at the same time; they *prefer* to do so (again I assign intention to what is only mathematics).

Why do the two classes of particle divide according to their social or antisocial "instincts"? The answer lies in a subtle yet relatively simple feature of quantum mechanics that I will discuss at the end of the chapter. It is a feature specific to quantum mechanics—a feature with no counterpart whatsoever, not even approximately, in classical physic, yet with the most far-reaching consequences in the large-scale world we inhabit.

## Fermions

The twenty-five-year-old Wolfgang Pauli postulated the exclusion principle in 1925. Pauli, an Austrian, had been educated in Germany and later settled in Switzerland, also spending a good deal of time in the United States. His early career was meteoric. In 1921 he had earned his Ph.D. degree at Munich and also published a definitive survey of relativity theory, astonishing Albert Einstein by his grasp of the subject. In 1926, a year after introducing the exclusion principle, he led the way in applying Werner Heisenberg's brand-new quantum theory to the atom. He proposed the neutrino in 1930, by which time he was all of thirty and a professor at Zurich. In his later years, he was famous for his ability to intimidate physicists who were lecturing on their research results; he would sit in the first row, shaking his head and scowling. My acquaintance with Pauli was indirect, through a group of young German researchers with whom I worked in 1955–56. They corresponded frequently with Pauli, sending him their latest thoughts on particle theory.

His first reaction to their letters was always the same: "Alles Quatsch" ("total nonsense"). After a second round of correspondence, he allowed that there was perhaps a little sense in their ideas. By the third round, he was congratulating them on their insight.

Pauli formulated the exclusion principle near the end of a dozen-year period of confusion and frustration for physicists. After the twenty-seven-year-old Niels Bohr provided a quantum theory of the hydrogen atom in 1913, physicists knew that Planck's constant, $h$, must play an essential role within the atom, and they assumed that the ideas Bohr had laid out—electrons occupy stationary states and make quantum jumps between these states, while emitting and absorbing photons—were probably correct for all atoms. But a real quantum theory was lacking until Werner Heisenberg in 1925 (at age twenty-three) and Erwin Schrödinger in 1926 (at the advanced age of thirty-eight) tied the loose ends together and created the quantum theory that has stood till this day. Pauli's exclusion principle was part of what launched this revolution.

Before discussing the implications of the exclusion principle, I need to define *state of motion* and *quantum number*. What does it mean to say that an electron is in a certain state of motion or that it has certain quantum numbers?

An automobile traveling due west at constant speed on an absolutely straight road can be said to be in a certain state of motion. This "state" is defined not by where the car is but by how it is moving—at what speed and in what direction. Another car moving at the same speed in the same direction on the same highway is in the *same* state of motion as the first car—even if they are far apart. To give another automotive example, an Indianapolis race car tearing around the track, varying its speed and acceleration in exactly the same way on each circuit, is in a certain state of motion. Two cars are in the same state of motion if they have identical patterns of speed and acceleration, whatever their separation might be. A satellite circling the Earth has a state of motion defined not by where it is at a certain time but by its energy and angular momentum. Another satellite following a long, skinny elliptical path might have the same energy as the first satellite but less angular momentum. It is in a different state of motion. So the state of motion of an object is a "global" property related to the totality of its motion, not to some specific part of the motion.

For an electron in an atom, a physicist can't follow the details of the motion; nature doesn't allow it. The only available information *is* global information. It's as if the Indianapolis race car is moving so fast that you can't make out anything other than a general blur. You know it's confined to the race track and you know it's moving with a certain average speed, but you don't know where it is at any time. The electron in a state of motion within an atom is likewise a blur, with some probability of being in one place and another probability of being in a different place.

But not everything about the electron is fuzzy. It may have a definite energy, a definite angular momentum, and a definite orientation of the axis of its orbital motion. This leads to the possibility of assigning numbers to define the state—a number for the particular energy of the state, a number for its angular momentum, and a number for the orientation of the angular momentum. Since these three physical quantities are *quantized*—meaning they can take only certain discrete values—the numbers that characterize the state are also quantized. Accordingly, they are called *quantum numbers*. For instance, the *principal quantum number*, $n$, is chosen to be 1 for the lowest-energy state, 2 for the next state, and so on. It tells where the state lies in the ladder of allowed energies. The *angular-momentum quantum number*, $\ell$, measuring the angular momentum in units of $\hbar$, can be zero or any positive integer. Finally, the *orientation quantum number*, $m$, can take on negative and positive values ranging from $-\ell$ *to* $+\ell$.

Bohr made do with the single quantum number $n$. Over the years that followed, physicists concluded that a full description of an electron's state of motion required the three quantum numbers defined above: $n$, $\ell$, and $m$. They also concluded that electrons occupy "shells" in atoms and do not all cluster together in the lowest-energy state. This conclusion was required to account for the periodic table. The first shell, it appeared, could hold two electrons, the next shell eight electrons, and the shell after that also eight electrons. But there was no clear rationale for these numbers—until Pauli came along. He made two contributions. First, he said, there is an exclusion principle at work that permits no more than one electron to occupy a given state of motion. Second, he said, an electron must have an additional *degree of freedom*, characterized by a fourth quantum number that can take on only two possible values (which, as a matter of convenience, could be chosen to be $+\frac{1}{2}$ and $-\frac{1}{2}$).

Since a state of motion is characterized by a set of quantum numbers, another way to express the exclusion principle is to say that every electron in an atom has a different set of quantum numbers. (This is the way that Pauli put it.) The formulation enabled Pauli to explain why the first shell contains two electrons and the second shell contains eight electrons. The two electrons in the lowest shell both have $n = 1$, $\ell = 0$, and $m = 0$, but have opposite values of the fourth quantum number: one has $+\frac{1}{2}$ and the other $-\frac{1}{2}$. In the second shell, with $n = 2$, are a group of two electrons and another group of six electrons—a total of eight. The two electrons have $\ell = 0$ and $m = 0$, one with each value of Pauli's new quantum number. The six electrons have $\ell = 1$, with three values of $m$ ($-1$, $0$, and $+1$) and two values of the new quantum number. The exclusion principle without the fourth quantum number would have predicted the contents of the first two shells to be 1 and 4 rather than 2 and 8. Pauli's new quantum number provided the needed doubling. It looked like physics and chemistry were merging, with individual electron quantum numbers accounting for the periodic table—an exciting development indeed.*

The exclusion principle was barely out of Pauli's mouth before two youngsters in the Netherlands, Samuel Goudsmit (twenty-two) and George Uhlenbeck (twenty-four), interpreted the fourth quantum num-

---

* The simplest application of the exclusion principle with electron spin predicts that the first three shells should contain 2, 8, and 18 electrons, respectively, rather than the observed 2, 8, and 8 (18 comes in the fourth shell). Understanding this difference came quickly and required no change in Pauli's ideas. It has to do with the various environments in which different electrons move, some close to the nucleus and feeling its full force, others farther away, "shielded" in part by electrons in inner shells. The result is a "distortion" of the energy-level pattern that fully accounts for the periodic table.

George Uhlenbeck (1900–1988), left, and Samuel Goudsmit (1902–1978), right, flanking their professor Hendrik Kramers (1894–1952) at a summer school in Ann Arbor, Michigan, ca. 1928. Photo courtesy of AIP Emilio Segrè Visual Archives, Goudsmit Collection.

ber.* It indicates, they said, that the electron is spinning, and that the spin has a value of $(\frac{1}{2})\hbar$. Pauli's two-valued quantum number then simply reflects the two possible orientations of a spin of this magnitude. It can point "up" or it can point "down." Spin found ready acceptance. It explained why the fourth quantum number could have only two values and also explained why, in an ordinary atom, the energy of a state doesn't depend on the value of that quantum number. Changing the di-

---

* Both Goudsmit and Uhlenbeck later moved to the United States. Uhlenbeck held professorships at Rockefeller University and the University of Michigan. Goudsmit became editor of *Physical Review*, the most prestigious physics journal in America. Before the end of the second World War in Europe, he headed the Alsos mission, whose task was to find out what progress toward an atomic bomb had been made by German scientists.

rection of the electron's spin should not affect the electron's energy. That's for an "ordinary" atom. If the atom is placed in a magnetic field, things change. Then the energy *does* show a dependence on the spin orientation. Analysis of spectra emitted by atoms in magnetic fields showed a pattern consistent with a spinning electron, whose energy changes slightly when the electron spin flips from "up" to "down."

In 1926, hard on the heels of the exclusion principle, the discovery of electron spin, and the new theory of quantum mechanics, Enrico Fermi in Italy and Paul Dirac in England (both were in their mid-twenties at the time) independently generalized Pauli's work. They investigated the "statistics" of particles—whether electrons or some other kind of particle—that obeyed the exclusion principle. This means they explored how such particles would behave when two or more of them were together, whether in an atom or in any other environment. They discovered what we now call *Fermi-Dirac statistics*. The particles they studied we now call *fermions*. Pauli later established mathematically that any particle having half-odd-integral spin ($\frac{1}{2}$, $\frac{3}{2}$, $\frac{5}{2}$, and so on) is a fermion.

Imagine a world in which electrons do not obey the exclusion principle. You can be sure it is a world of imagination only, for in such a world there would be no living things and only the dullest kind of chemistry. Hydrogen and helium, the first two elements in the periodic table, would be pretty much the same as in our actual world. But lithium, the third element, would be even more stable, less chemically active, than helium. The three electrons in a lithium atom would all fall into the same lowest-energy state (with $n = 1$); whereas in the real lithium atom the third electron must occupy a higher level, an arrangement that accounts for the vigorous chemical activity of lithium. Moving along in the periodic table, every element would have all of its atomic electrons clustered together in the first shell. Each element would be even more inert than the one before it.

Now imagine another world, in which electrons *do* obey the exclusion principle but are spinless. This might be an interesting and colorful world, but it would be very different from the one we inhabit. The first few electron shells in this world would hold 1, 4, and 4 electrons, respectively. Element number 2, helium, would have only one electron in

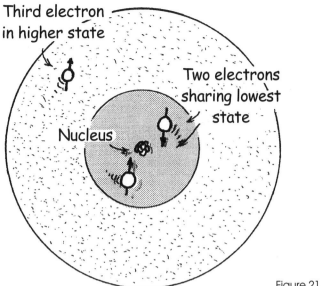

Third electron in higher state

Two electrons sharing lowest state

Nucleus

Figure 21. Lithium atom.

its first shell. Its other electron would have to find a home in a higher-energy state. Instead of being the noble gas we know (that is, one which does not combine with other elements), helium would be chemically active. The first noble gas would be boron, element number 5. The periodic table in this hypothetical world would have twice as many periods as the periodic table in our world. Who knows what interesting chemistry might result?

I mention these two fanciful worlds only as a way of emphasizing how remarkable (and wonderful) it is that electrons do have spin and do obey the exclusion principle. I can think of no more marvelous link in all of physics than that between the few quantum numbers that describe states of motion in atoms and the whole panoply of chemistry and life that springs from the periodic table. Life-nurturing oxygen behaves as it does because the eight electrons in each oxygen atom align their orbital and spin angular momenta only in certain ways to assure that no two electrons occupy the same state of motion. The carbon atom, backbone of living matter, reaches out to bond with numerous other atoms through electron sharing because its second shell is exactly half full, containing four out of a possible eight electrons. That number eight is the number of ways that electrons (or, for that matter, any fermions) with either zero or one unit of orbital angular momentum and one-half unit of spin angular momentum can arrange themselves without duplicating any quantum numbers. The marvel that is the periodic table finds its explanation in the marvel of spinning electrons that obey the exclusion principle.

It is not only within atoms that the exclusion principle has practical consequences. It is also within atomic nuclei, where neutrons and protons, being fermions, must avoid sharing states of motion, and within metals, where great numbers of electrons spread themselves over large distances but still respectfully avoid duplicate occupation of any energy state.

## Nuclei

Within a nucleus, two protons cannot share the same set of quantum numbers, and neither can two neutrons; but there is no bar to a proton and a neutron occupying the same state of motion. The exclusion principle applies only to *identical* fermions. The inside of a nucleus is a bit like a coed dorm in which boys avoid one another but don't mind studying with girls, while girls avoid one another but don't mind studying with boys. The exclusion principle helps to explain why the fusion process that powers the Sun and other stars releases so much energy. The nucleus of helium 4 that results from the fusion of hydrogen is particularly stable. Its two neutrons and two protons all occupy the lowest-energy state—they are all in the first shell. These four particles, in coming together, release a great deal of energy as they fall into the lowest-energy state within the nucleus. So the sun shines brightly.

Not until 1948 did physicists realize that a shell structure exists within nuclei just as it does within atoms. They had assumed that the protons and neutrons within a nucleus milled around elbow to elbow like molecules in a liquid, rather than orbiting like planets. Interestingly, this *liquid droplet model* of the nucleus served adequately to provide a theory of nuclear fission—even to predict that the nucleus of the still-undiscovered plutonium 239, like that of uranium 235, would be susceptible to fission with slow neutrons. Soon, however, evidence accumulated for *closed shells*, certain numbers of neutrons or protons that filled a

Figure 22. Nucleus of a helium atom.

given energy level, requiring that additional nuclear particles take up residence in a higher-energy state. Because these closed-shell numbers were initially rather mysterious, they came to be called *magic numbers* (a term possibly coined by the Princeton physicist Eugene Wigner). Maria Mayer* in Chicago and, independently, Hans Jensen in Heidelberg developed the shell model of the nucleus, which recognized that despite some liquid-like properties, nuclei also showed evidence of orbiting fermions that scrupulously obeyed the exclusion principle.

Two of the magic (closed-shell) numbers happen to be 82 and 126. By chance, the nucleus of lead 208 contains 82 protons and 126 neutrons. It is doubly magic, and as a result is particularly stable. It falls just one particle short of being the heaviest stable (nonradioactive) nucleus. That honor belongs to bismuth 209, containing 83 protons and 126 neutrons. All heavier nuclei are radioactive. In current searches for much heavier elements, physicists hope to verify that there exists an "island of stability" (which should actually be called a "not-quite-so-unstable island") at 114 protons, an expectation based on the theory of energy levels within nuclei. Indeed, the currently heaviest known element

---

* For many years, Maria Mayer had to content herself with lower-level teaching and research jobs while her chemist husband, Joseph Mayer, enjoyed professorships. When I went to see her in the 1950s to discuss nuclear research, she held two half-time positions, one at Argonne National Laboratory and the other at the University of Chicago. Eventually, both she and her husband were named professors at the University of California, San Diego. She received the Nobel Prize in 1963.

is number 114 (not yet named), discovered in Dubna, Russia, in 1999. Its "long" lifetime of thirty seconds supports the idea of greater stability at this proton number. (The next-heaviest known element, number 112—which was discovered in Darmstadt, Germany,* in 1996—lives for only 280 $\mu$s: less than a third of a millisecond.) The lack of success in finding any element heavier than number 114, despite great efforts, also suggests that there is something special about 114, further evidence of the way in which the exclusion principle governs the behavior of the nucleons (neutrons and protons) within the nucleus.†

"Shell" is not really a very good term to describe a set of states of motion with nearly the same energy, because an electron or proton or neutron in a given energy state doesn't spread itself along a thin shell. The particle's cloud of probability spreads over a volume, and different shells overlap, sharing some of the same space. The energy and angular momentum in a given state of motion are sharply defined, but the particle's position is not. This is particularly true of the crowded nucleons within a nucleus.

A better visual model than a set of shells might be an apartment house of many stories, with occupancy on each floor limited (imagine signs specifying maximum occupancy, like those in restaurants and other public spaces). The apartment dwellers, being good fermions, will strictly obey the occupancy limits. Moreover, as fermions, their behavior will differ somewhat in summer and winter. In the coldest weather, they will go not a bit higher than they have to. As many of them as possible will crowd onto the lowest floors. A person will go higher only when forced to by full occupancy below. In hot summer weather, by

---

* Discoveries in Dubna and Darmstadt have been recognized in the names of three heavy elements, dubnium (105), darmstadtium (110), and hassium (108). (Hassias is the Latin name for Hesse, the German state in which Darmstadt is located.) No less than four element names recognize early contributions by researchers in Berkeley, California: berkelium (97), californium (98), lawrencium (103), and seaborgium (106), the latter two named for Ernest O. Lawrence and Glenn T. Seaborg.

† In 2002 a team of scientists claimed to have discovered elements 116 and 118, but the claim was withdrawn in the face of evidence that some of the supporting data had been fabricated—a regrettable and rare, but not wholly unknown, phenomenon in physics.

contrast, although there will still be a general preference for lower floors, some people will seek out higher floors even when lower floors are not fully occupied. If the apartment dwellers choose to follow the statistics of Fermi and Dirac, we physicists could predict exactly how many of them would be found on each floor in every season.

## Metals

The apartment-house model may sound a bit far-fetched, but it almost perfectly describes electrons in a metal. At a temperature near absolute zero (the "depths of winter"), the electrons fill up the lowest-energy states one after the other, only as high as necessary to accommodate all the electrons. At high temperatures ("summer"), some of the higher-energy states are only partly filled and a small fraction of the electrons move about in still higher-energy states. (There is one difference between the apartment house and the metal. In the metal, gaps can arise in the energy-level pattern. It's as if some floors in the apartment building were simply missing, with air—and presumably girders—between the lower and higher floors. No people there, just space.)

The apartment-house model is also roughly appropriate for atoms and nuclei. The difference is that because of the large energy difference between permitted states in an atom, and the even larger difference between energy states in the nucleus, what constitutes a low temperature and a high temperature is quite different for these structures than for a metal, which has very closely spaced energy states. For the atom, room temperature is pretty much equivalent to absolute zero. It takes a temperature of thousands of degrees to cause some of the electrons in an atom to jump spontaneously to higher-energy states, and a temperature of millions of degrees for similar excitation in a nucleus.

## Bosons

In 1924, Satyendra Nath Bose ("BO-suh"), a thirty-year-old physics professor at the University of Dacca,* sent off a letter to Albert Einstein

---

* Dacca, now the capital of Bangladesh, was then in India.

in Berlin. Enclosed was a paper of his entitled "Planck's Law and the Hypothesis of Light Quanta," which had been rejected by a leading British journal, *Philosophical Magazine*. Undaunted by the rejection, Bose had decided to approach the world's most celebrated physicist, emboldened perhaps by the fact that he himself had translated Einstein's relativity text from German to English for the Indian market, or perhaps simply because he was quite confident that his paper had something significant to say.

The "Planck's Law" in the title of Bose's paper is the mathematical law introduced by Max Planck in 1900 that gives the distribution of energy among different frequencies of radiation within an enclosure at constant temperature—the so-called *black-body radiation*, or *cavity radiation*, that I discussed in Chapter 5. Recall that in Planck's hands the formula $E = hf$ referred not to the energy of a photon of frequency $f$ but to the minimum energy that a material object could give to or take from radiation. Planck assumed, as did most physicists after him for nearly a quarter of a century, that what was quantized was energy transfer to and from radiation, not the radiation itself. This despite the fact that Einstein had proposed the photon in 1905 (actually so named only much later)* and that Arthur Compton had seen evidence for photon-electron scattering in 1923. When Bose wrote his paper in 1924, he still referred to "light quanta" (photons) as *hypothetical* entities. His paper would play a major role in transforming the photon from hypothesis to accepted reality.

In the paper he sent to Einstein, Bose showed that he could derive Planck's law by postulating that the radiation consists of a "gas" of "light quanta" that do not interact with one another and which can occupy any energy state irrespective of whether another "light quantum" is already in that state.† Einstein at once realized that Bose's derivation was a giant step forward from Planck's original derivation and that it provided strong, if indirect, evidence for the reality of the light quantum. Einstein personally translated Bose's paper from English into German

---

* Gilbert Lewis coined the term "photon" in 1926.

† Bose had no inkling of the exclusion principle, which was advanced in the following year and was applied by Pauli to electrons, not photons.

and forwarded it to a leading German journal, *Zeitschrift für Physik*, with his recommendation that it be published. It was, promptly.

Einstein, intrigued by Bose's work, temporarily turned his attention away from his effort to unify electromagnetism and gravitation and considered how atoms would behave if they followed the same rules as photons. His paper was published hard on the heels of Bose's paper in the same year. So came into being what we now call *Bose-Einstein statistics*. A few years later, Paul Dirac suggested that particles obeying these statistical rules be named *bosons*.*

One thing that Einstein worked out was what would happen to a gas of atoms at extremely low temperature. (He assumed that his atoms satisfied Bose-Einstein statistics, which, as it turns out, about half of all atoms do.) Let me go back to the apartment-house model. For boson residents, there is no limit to how many may occupy the first floor or on any other floor. You might think this would make all the bosons in a given assemblage gather in the lowest-energy state. They do have some *tendency* to do so, but it is a tendency that is fully realized only at extremely low temperatures. In "mild weather," many bosons gather on the first floor, but many others spread themselves over upper floors. Only when the temperature plunges below Antarctic levels (well, actually to within less than a millionth of a degree of absolute zero) do all the bosons cluster on the first floor, the lowest-energy state. Einstein realized (to push the apartment-house model a bit further) that under this circumstance, the bosons would not only occupy the same floor—that is, have the same energy—but would distribute themselves identically over that floor. Every boson in the collection would be in identically the same state of motion as every other one, so that they would be totally overlapping and interpenetrating. Each one would occupy the *whole* floor. Every atom spreads itself out with a probability distribution the same as that of every other atom (now we call it a *Bose-Einstein condensate*). It took seventy years for experimenters to catch up with theory and produce a Bose-Einstein condensate in the lab. The principal reason for the delay was the great difficulty in pushing temperature to the extraordi-

---

* Dirac, a famously modest man, also named *fermions*, whose properties both he and Fermi discovered.

narily low value required. Neither Bose nor Einstein lived to see the verification of this remarkable behavior of bosons.

One difference between fermions and bosons has to do with their numbers. Evidence suggests that the total number of fermions in the universe is constant (provided antifermions are assigned negative particle number),* whereas the number of bosons can change. The fermion rule comes into play in every individual particle reaction. For instance, in the decay of a negative muon into an electron, a neutrino, and an antineutrino,

$$\mu^- \rightarrow e^- + \nu_\mu + \bar{\nu}_e,$$

there is one fermion present before the decay and one afterward (assigning negative particle number to the antineutrino):

$$1 \rightarrow 1 + 1 + (-1).$$

Similarly, in the decay of a neutron,

$$n \rightarrow p + e^- + \bar{\nu}_e,$$

there is one fermion before the decay and a net of one afterward. Again,

$$1 \rightarrow 1 + 1 + (-1).$$

When an electron and positron annihilate to create a pair of photons, the count of fermions is zero before and after:

$$e^- + e^+ \rightarrow 2\gamma,$$
$$1 + (-1) \rightarrow 0.$$

This is an example in which the boson count changes, in this case going from zero to two. Similarly, when a proton collides with another proton in an accelerator, various bosons may be created. Here is an example:

---

* Black holes may provide an exception to this rule. Theory says that a black hole that has gobbled up a batch of fermions does not preserve their number—indeed, that the number of fermions in a black hole is a meaningless concept.

$$p + p \rightarrow p + n + \pi^+ + \pi^+ + \pi^-.$$

Three bosons emerge where there were none before. (Note that the number of fermions, two, is preserved.) To give just one more example, in the decay of a negative pion into a muon and an antineutrino,

$$\pi^- \rightarrow \mu^- + \bar{\nu}_\mu,$$

the boson count goes from one to zero, and the fermion count from zero to zero (again assigning negative particle number to the antineutrino). No one has an answer to the deep questions: Why do fermions preserve their numbers?* Why do bosons come and go in arbitrary numbers? Why do black holes flout the rules?

## Bose-Einstein Condensate

In 1995, Eric Cornell and Carl Wieman, working at the Joint Institute of Laboratory Astrophysics in Boulder, Colorado, were the first to create and study a Bose-Einstein condensate.† Their initial success came with a few thousand atoms of rubidium, cooled to 200 billionths of a degree above absolute zero. At this temperature, the rubidium atoms are ambling along at a speed of about 8 millimeters per second, or 90 feet per hour (compared with their room-temperature speed of about 300 meters per second, or 650 miles per hour). Cooling and slowing are really the same thing, for an atom's average speed measures its temperature. To achieve the 200-nanokelvin temperature, a record at the time, Wieman and Cornell used *laser cooling* and *magnetic trapping*. Laser light slows the atoms to a walk; then magnetic fields corral them in a small region while evaporative cooling—the same phenomenon that causes you to shiver in a wet bathing suit—chills the atoms even further.

Rubidium is element number 37 in the periodic table, so its nucleus contains 37 protons. Surrounding the nucleus are 37 electrons. That's a

---

* Theory *does* answer the question: Why do antiparticles have to be counted as negative particles? This is not a mystery.

† For this achievement, they shared the 2001 Nobel Prize in Physics with Wolfgang Ketterle of MIT, another pioneer in the field.

Carl Wieman (b. 1951), left, and Eric Cornell (b. 1961) in Boulder, Colorado, 1996. Photo by Ken Abbott; courtesy of the University of Colorado, Boulder.

total of 74 fermions so far. If the nucleus contains an even number of neutrons, the grand total number of fermions is even and the atom is a boson. This is true of two common isotopes of the element, rubidium 85 (with 48 neutrons) and rubidium 87 (with 50 neutrons). (You could count quarks and electrons and reach the same conclusion: that these two isotopes of rubidium are bosons.) At MIT, Wolfgang Ketterle achieved a Bose-Einstein condensate with larger numbers of sodium atoms. As mentioned earlier, an atom of sodium contains 11 protons, 12 neutrons, and 11 electrons, so it, too, is a boson.

To help you visualize a Bose-Einstein condensate, I will return to the apartment-house model. The year is 2126. Gathered in a first-floor apartment are 85 clones, genetically identical individuals. The apartment is so large that they don't get in each other's way. Looking in from outside, you can see them as individuals, even if you can't tell who is who. As the cryogenic thermostat is lowered to a setting of 200 nanokelvins, each individual is transformed into a murky cloud that spreads

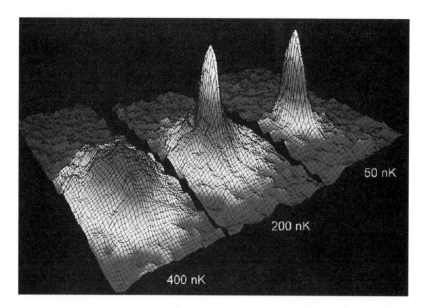

Data map by Wieman and Cornell showing the velocity distribution in a cloud of rubidium atoms as the temperature is lowered from 400 to 200 and then to 50 nanokelvin (billionths of a degree above absolute zero). The peaks that appear when the temperature is 200 nK or lower indicate the formation of a Bose-Einstein condensate in which the atoms move at near-zero velocity. Image courtesy of Mike Matthews, Carl Wieman, and Eric Cornell, University of Colorado, Boulder.

throughout the apartment (no trick at all for a Hollywood special-effects person). Each cloud overlaps identically with every other one, so the apartment is filled with a single dense cloud. Watching from outside, you have no way of picking out individuals. It looks like the apartment has been leased to a single great blob. Yet the individuality of the clones is not lost. When the thermostat is turned back up to a millionth of a degree or so, the cloud separates back into 85 individuals, none the worse for their experience.

Already intriguing to physicists is the practical question: Will Bose-Einstein condensates find useful application? If history is any guide, the answer is likely to be yes. Whenever scientists have understood and gained control over some new form of matter, they have found ways to use the understanding and the control for useful purposes. Condensates may, for instance, find use in precision measurements of fundamental constants, or in quantum computers, or in a new kind of laser that uses a beam of atoms instead of a beam of light.

## Why Fermions and Bosons?

How does Nature contrive that every particle is social or antisocial—either wanting to cluster in the same state of motion or refusing to do so? Classical theory cannot provide an answer to this question, not even an approximate answer. That's why it's such an interesting question. To give an answer, I have to turn to a feature of quantum theory that is mathematical, but (I hope) comprehensible. The answer depends also on the fact that there exist in nature particles that are identical.

The exclusion principle does *not* say that no two fermions can occupy the same state of motion. It says that no two fermions *of the same kind* (two electrons, or two protons, or two red up quarks) can occupy the same state of motion. Similarly, it is only two bosons *of the same kind* (two photons, or two positive pions, or two negative kaons) that prefer to occupy the same state of motion. If every particle in the universe differed from every other one, in no matter how small a particular, it wouldn't matter much whether a particle was a fermion or a boson, for then there would be no bar to fellow fermions occupying the same state of motion, and no incentive for fellow bosons to do so. Particles would then be like baseballs—all slightly different and with no incentives or disincentives to aggregating. So the fact that in the subatomic realm we find entities that are truly identical has, literally, cosmic consequences. If, for instance, electrons were not all precisely identical, they would not fill sequential shells in atoms, there would be no periodic table, and there would be no you and no me.

One thing that sets quantum theory apart from theories that went before is the fact that it deals with *unobservable* quantities. One of the unobservables is called the *wave function*, or *wave amplitude*. The *probability* that a particle is in a certain place or is moving in a certain way is proportional to the *square* of the wave function. So the wave function multiplied by itself relates to what is observable, but the wave function alone does not. This means that whether the wave function is positive or negative has no observable consequences, since the squares of both positive and negative numbers are positive.* Every once in a while in this book, I have to ask you to fasten your seat belt (or step out and skateboard to the next milepost). This is such a time.

If particle number 1 is in state A and particle number 2 is in state B, the wave function for the combination of the two particles can be expressed by the notation A(1)B(2). If, instead, it is particle number 2 that is in state A and particle number 1 that is in state B, the notation would be A(2)B(1). Now here comes the issue of indistinguishability. If particles 1 and 2 are truly identical, there is no way of knowing which one is in state A and which one is in state B. So the two combinations above both describe the same physical situation: one particle in state A and one in state B. Early quantum theorists discovered that one way to cope with this "undecidability" about which particle is where is to add together the two wave functions:

A(1)B(2) + A(2)B(1).

This is a satisfying solution to the dilemma. It says, in effect, that both particles occupy both states, with each having a 50-percent chance to be in either state. If you perform the mathematical maneuver of interchanging the 1's and the 2's in the above expression, what you get after the interchange is the same as what you had before. This is consistent with the physical situation: interchanging the two particles has no observable consequence, because the two particles are identical.

Quantum theory permits a different solution to the dilemma (the dilemma of dealing with identical particles that can switch places with no observable consequences). One can insert a negative instead of a positive sign between the wave functions that are combined:

A(1)B(2) − A(2)B(1).

---

* The full story is a bit more complicated than this. The unobservable wave function can be a *complex number*, meaning a combination of a real and an imaginary number. That moves it even further from observability than it would be if merely negative. Something called the *absolute square* of a complex number is a positive quantity, and that's what is observable.

Now interchanging the 1's and the 2's produces

$$A(2)B(1) - A(1)B(2),$$

which is the negative of what you started with. But this is all right, because it is only the *square* of this quantity that has observable meaning, and the square is unchanged.

The first combination displayed above, the one with the plus sign, is called a *symmetric wave function*. The second combination, the one with the minus sign, is called an *antisymmetric wave function*. It turns out that the symmetric wave function describes two identical *bosons* and the antisymmetric wave function describes two identical *fermions*. Theorists have tried out combinations other than just addition and subtraction, but Nature seems to have settled on just these two. Every known particle is a boson or a fermion.

The big difference between bosons and fermions shows up when the two identical particles are in the *same* state of motion. If both particles are in the state A, the symmetric combination becomes

$$A(1)A(2) + A(2)A(1),$$

which is just $2A(1)A(2)$. No problem. That combination describes a pair of bosons in the same state. But what if two fermions are—or would like to be—in the same state A? The antisymmetric wave function for two identical particles in the same state is

$$A(1)A(2) - A(2)A(1).$$

This is zero! The two terms cancel. Two fermions *can't* be in the same state. This is the startlingly simple (although admittedly subtle) mathematical reason for the exclusion principle.

It is almost scary to think of the consequences—from Bose-Einstein condensates to the entire periodic table of the elements—that flow from merely a plus sign or a minus sign. These two choices rest, in turn, on two facts: that quantum theory deals with *unobservable* wave functions and that particles of a given type are clones of one another, truly *identical*. It is no wonder that physicists stand in awe of the power of mathematics to describe Nature.

## chapter 8

# Clinging to Constancy

"All is change." So said Euripides in the fifth century B.C. You might be inclined to agree with him. Nothing you encounter in the world around you is permanent and unchanging. Writing even before Euripides, Heraclitus put it this way: "Nothing is permanent but change." Physicists don't dispute the evidence of the senses that everything around us changes ceaselessly. But they have identified some things in nature that *do* remain constant. Such quantities are said to be "conserved." A law that says one quantity remains constant while others change is called a *conservation law*.

I touched on various conservation laws in earlier chapters. You know that energy (including mass) is conserved; also momentum, electric charge, quark and lepton flavors, and baryon number. That short list includes most of the important conservation laws. In this chapter I will add a few more to the list, and will distinguish between conservation laws that are absolute (so far as we know) and those that are partial. And I will examine the interesting question: Why have conservation laws moved from playing bit parts to holding starring roles in the script that physicists write for Nature?

Conservation laws are relative newcomers in the history of physics.

Aristotle, like Euripides and Heraclitus before him, focused principally on change. So did most scientists until the modern era. Johannes Kepler, with his second law of planetary motion, may have introduced the first conservation law, at the beginning of the seventeenth century. This law says that a hypothetical line drawn from the Sun to a planet sweeps out equal areas in equal times. The planet moves faster when it comes closer to the Sun and slower when it is farther from the Sun, always varying its speed in just such a way that the rate at which the radial line is sweeping out area is constant. Now we understand Kepler's second law to be a consequence of angular-momentum conservation. As each planet pursues its orbit, some of its features vary—its speed, its direction of motion, its distance from the Sun—but its orbital angular momentum remains constant. The Earth's reliable turn about its own axis once every twenty-four hours is also a consequence of angular-momentum conservation, in this case spin angular momentum.

Not long after Kepler's work, Galileo Galilei recognized that an object free of outside influences moves indefinitely at constant velocity—a kind of conservation law. Later in the seventeenth century, Christiaan Huygens and Isaac Newton introduced momentum conservation, although Newton, in his great work on motion, still focused mainly on change. Benjamin Franklin's proposal of charge conservation came in the eighteenth century. The most dramatic step toward assigning primacy to conservation laws came in the mid-nineteenth century, when scientists proposed the general law of energy conservation, encompassing both mechanical energy and heat energy. At the time, chemists were relying on the law of mass conservation. This law crumbled in Einstein's hands. With his famous formula $E = mc^2$, Einstein showed that mass is just congealed energy and must be incorporated into a larger law of energy conservation. (In chemical reactions, mass changes are imperceptible, so mass conservation remains a useful tool for chemists. In subatomic particle reactions, on the other hand, mass changes can be dramatically large—consider the annihilation of a proton and an antiproton, in which mass disappears entirely.) Finally, the laws of conservation of flavor, color, and particle type were added in the middle and late twentieth century.

Why are conservation laws now so central in our description of na-

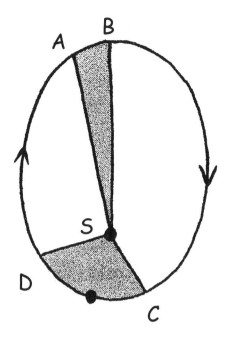

Figure 23. Kepler's second law: A planet sweeps equal area in equal time.

ture? One reason is the quite human tendency to see beauty in simplicity. What idea could be simpler—and more powerful—than the idea of constancy amid the incessant change in the world around us? Another reason is that conservation laws are connected to principles of symmetry. I will say more about this fascinating link later in the chapter. Still another reason is a very practical one. In the maelstrom of a particle reaction, there is no hope (not even in principle) of discovering what is going on, instant by instant. We have to settle for a before-and-after view of the reaction. The physicist knows what is there before the reaction and can measure what is there afterward. The details in between are shielded from view. Theory suggests that many different things are happening at once, all higgledy-piggledy on top of one another. What brings order are conservation laws. Some quantities survive unscathed, the same after as before.

A classical law of physics, a law of change, can be called a *law of compulsion*. It describes what happens, indeed what *must* happen, under particular circumstances. If Zeus places a planet at a certain distance from the Sun and nudges it in a certain direction (that is, supplies the *initial conditions*), Newton's law of motion takes over and tells where that planet will be—indeed, where it must be—forever afterward, as a result of the forces of the Sun and other planets acting on it. A law of compulsion is normally expressed by a mathematical equation. For instance, consider the fall of a marble that you have released from rest. The equation that states just how far it has fallen after some lapse of time is $d = (\frac{1}{2})gt^2$. An equation is a recipe. This one says: Take the time of fall in seconds. Square it. Multiply by the acceleration of gravity, $g$ (9.8 meters per second squared). Multiply that result in turn by $\frac{1}{2}$ and you have the distance of fall in meters. The equation gives you the instant-by-instant information on where the marble must be on its way to the floor. No such detailed information is available in a particle reaction.

A conservation law, by contrast, is a *law of prohibition*. It allows many possible things to happen, but prohibits all chains of events that violate the law. For example, if a high-energy proton in an accelerator strikes another proton, there are many possible outcomes. A few of dozens of possibilities might be

$$p + p \rightarrow p + p + \pi^+ + \pi^- + \pi^\circ,$$
$$p + p \rightarrow p + n + \pi^+ + \pi^\circ,$$
$$p + p \rightarrow p + \Lambda^\circ + K^\circ + \pi^+.$$

But it's not "anything goes." Conservation laws tell us that the particles spraying from the collision must have a total momentum equal to the momentum of the initial proton; that the total charge of the products must be $+2$; that the baryon number of the products must add to 2; that there must be a net count of 2 fermions in the products; and that the mass energy plus kinetic energy of the products must equal the original energy (mass + kinetic). This last requirement limits the total number of massive particles that can be created, for each one soaks up some of the available energy.

Connected with the *prohibition* of the conservation law is a quantum *permissiveness*. The conservation law itself doesn't have anything to say about which of the theoretically possible reactions (such as the three displayed above) will actually occur. But physicists have reason to believe that *every* possible reaction will in fact occur, sooner or later, with some probability. (That's why I call it *quantum* permissiveness. It is quantum probability that opens up all the possibilities.) So if protons of a given energy are fired one after another at a target containing protons, every outcome that is allowed by the conservation laws will eventually occur—some often, some not so often, some perhaps only quite rarely. This permissiveness is a striking adjunct to the prohibitions of conservation laws. It means not merely that everything not forbidden by one or more conservation laws is *allowed* (which is pretty much the way laws of prohibition work in human affairs), but that everything not forbidden by the conservation laws is *obligatory* (imagine that because of construction you are prohibited from following your normal route from your home to your office, and accordingly must, one day or another, follow every other possible route to work).*

## Invariance Principles

Conservation laws express one kind of constancy in the physical world: the constancy of certain physical quantities (energy, momentum, and so on). There is another kind of constancy, equally striking and significant: the constancy of the physical laws themselves. An *invariance principle* is a statement about such constancy. It says that the laws of physics remain unchanged as experimental conditions change in some specified way. An example is *location invariance*: the laws of physics are the same in one place as in another. If an experiment works in a certain way in Batavia, Illinois, it will work in the same way and give the same results when conducted in Geneva, Switzerland. This seems obvious, but only because we are so used to experiencing no surprises in the way nature behaves as we move from one place to another. It is actually a deep truth about nature. It illustrates the homogeneity of space. Every point in space is equivalent to every other. (Such a statement is an example of a *symmetry principle*, which I will discuss later in the chapter.) Would

---

* To make this example closer in spirit to the world of physics, imagine that you are also prohibited from arriving late to work. Then the number of routes available to you would be limited, and you could follow only those that got you to work on time.

physics be possible if the results of an experiment depended on where the experiment was carried out? Yes, but physics would be a great deal less simple. Physicists would have to figure out not only the laws of physics, but the rules about how those laws changed from one place to another. In making space homogeneous and providing for location-invariance, nature has been kind to us.

Another invariance principle that we take for granted is *direction invariance*. The result of an experiment does not change if the apparatus is rotated. The outcome of a proton-proton collision, for example, doesn't depend on whether the projectile proton is flying eastward or flying northward before the collision. Again, this is an idea that seems obvious, and for the same reason. Our everyday experience with the world around us conditions us to believe it. But it's not "necessary." It just happens to be a feature of the universe we live in. One could imagine hypothetical universes with preferred directions—universes in which the laws of nature depended on direction. In our universe, lacking any preferred direction, space is said to be *isotropic*—the same in every direction.

Albert Einstein, in his 1905 theory of special relativity, introduced another invariance, which is called *Lorentz invariance* (it is named for Hendrik Lorentz, a noted Dutch physicist whose mathematical formulas Einstein used). Lorentz invariance states that the laws of nature which hold in an inertial frame of reference hold also in a frame of reference moving uniformly with respect to that frame. To appreciate the meaning of this statement, you must first understand what an *inertial frame of reference* is. It is a frame of reference in which objects free of outside influences are unaccelerated. An astronaut floating in an orbiting spacecraft is in an inertial frame; he drifts across the cabin without acceleration. You, to good approximation, are also in an inertial frame when you are standing still or when you are driving down a straight highway at constant speed. (On the other hand, when a car in which you are riding makes a turn and you are "thrown" to one side, you are in a noninertial frame.)

Although not aware of it, you have probably put Lorentz invariance to the test many times. If you sip a beverage or munch a snack while driving down the highway and don't spill a drop or a crumb, it may be because the laws of nature are the same in your moving car as in your car when it is parked. If you are flying in an airplane (in still air) and drop a coin, you see it fall to the floor exactly as if the plane were parked at the gate, because the laws of nature in the moving plane are the same as in the parked plane. (If you took the trouble to bring some precision measuring apparatus on board, you would find that the fall of the coin in the moving plane is, in every detail, exactly the same as in the parked plane.) Nearly three hundred years before Einstein's work, Galileo knew that the laws of mechanics were the same in a uniformly moving frame as in a stationary frame. Einstein's genius (you might say his bravery) was to propose that not just the laws of mechanics, but *all* laws, are the same in frames of reference moving uniformly with respect to one another. It is when the principle of Lorentz invariance is extended to the laws of electromagnetism that the astonishing consequences of relativity theory, such as time dilation and space contraction, are revealed.

You can see now that conservation laws and invariance principles, although different ideas, have common features. Both describe something that remains constant while other things change. Both, accordingly, bring simplicity to the description of nature. For an invariance principle, it is the laws of nature that remain unchanged; for a conservation law, it is one specific physical quantity that remains unchanged. The change referred to by an invariance principle is change of conditions under which an experiment is performed; the change referred to by a conservation law is the physical change that occurs as a process unfolds. Notice also the difference between what is general and what is specific. The invariance principle: *all* laws remain unchanged for *one* particular change of conditions. The conservation law: *one* quantity remains unchanged for *all* possible physical processes.

## Absolute Conservation Laws and Invariance Principles

The "big four" conserved quantities of the large-scale world—energy, momentum, angular momentum, and charge—are also conserved in the subatomic world. This is no surprise, for everything in the large-scale world is built ultimately of subatomic units. So you can think of the causal link going from small to large: energy, momentum, angular momentum, and charge are conserved in the large-scale world *because* they are conserved in the subatomic world. The conservation laws that govern these quantities are regarded as absolute. An *absolute conservation law* is one for which no confirmed violation has ever been seen and which is believed to be valid under all circumstances. Moreover, we have theoretical reason to believe that these four laws are absolute. Relativity and quantum theory join to predict that these laws *should* be valid. But experiment is the final arbiter. No amount of beautiful theory trumps experiment. Calling these conservation laws absolute must be as tentative as every other firm pronouncement about nature.

## Energy

In Chapter 6, I introduced the "downhill rule" for quantum jumps. A spontaneous transition must be from a higher to a lower-energy state. A particle can decay only into particles whose total mass is less than the mass of the decaying particle. And, as described earlier in this chapter, energy conservation also plays an essential role in "uphill" events such as the creation of new particles in a proton-proton collision. There is such abundant evidence that the energy after a reaction is the same as the energy before that the law of energy conservation becomes a practical tool for analyzing the complex debris created in particle collision.

## Momentum

Momentum conservation is so well established that it, like energy conservation, becomes a tool of analysis in particle processes. With the help of these laws, physicists can work backward to deduce the masses of new particles that are created.

One of the prohibitions resulting from the joint workings of energy conservation and momentum conservation is the prevention of one-particle decays. That is, an unstable particle cannot decay into only one other particle, even if the downhill rule is obeyed. Consider, for example, the hypothetical decay of a lambda particle into a neutron and nothing else:

$$\Lambda^\circ \not\rightarrow n.$$

The slash through the arrow means that the decay does not occur. If it did occur, charge, baryon number, and energy (it is downhill in mass) would all be conserved, but momentum would not. Suppose that the lambda is initially at rest. Some of its mass energy would appear as the mass energy of the neutron, and some would go into kinetic energy of the neutron. To conserve total energy, the neutron would have to be propelled away. But then momentum would not be preserved, for momentum would be zero before the decay and nonzero afterward. What if the neutron that is created *doesn't* fly away? That takes care of momentum conservation (it is zero before and after), but then energy is not conserved (it is less after than before). So the two conservation laws taken together prohibit one-particle decay.

The argument above applies to a lambda particle that is at rest before it decays. What if it is moving and therefore possesses kinetic energy as well as mass energy? Does that enable it to decay into a single particle? No. The argument is pretty much the same as the one presented in the discussion of the downhill rule in Chapter 6. If the decay doesn't occur in a frame of reference moving with the initial particle, it doesn't occur in any frame.

## Angular Momentum

An oddity of the quantum theory of angular momentum is that although angular momenta that are integers (in units of $\hbar$) can combine to make total angular momenta that are also integers, angular momenta that are half-odd integers combine differently depending on whether there are an even or odd number of them. An odd number combine to make a total angular momentum that is half-odd-integral, while an even number of them combine to make a total angular momentum that is integral. This recital may make your head spin. Here are some examples to clarify it. A neutral pion can conserve angular momentum while decaying into two photons:

$$\pi^\circ \to 2\gamma.$$

The pion's angular momentum is zero. Each photon has one unit of angular momentum. To see that angular momentum is conserved (zero before and after), you need only visualize the two photons spinning oppositely so that their angular momenta, as vectors, add to zero.

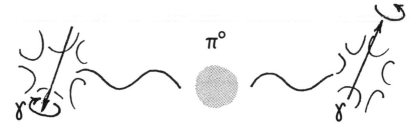

Figure 24. Decay of neutral pion.

Next consider the annihilation of an electron and a positron to create two photons:

$$e^- + e^+ \rightarrow 2\gamma.$$

In this example, there is an even number of fermions before the reaction, so the final product must have integral angular momentum. You can visualize the two electrons spinning oppositely, to have total angular momentum zero; the two final photons must then also spin oppositely to preserve zero angular momentum.

When a neutron decays into a proton, an electron, and an antineutrino, it is hemmed in by the strictures of several conservation laws, including that of angular momentum. In the decay, represented by

$$n \rightarrow p + e^- + \bar{\nu}_e,$$

one fermion goes to three (actually to two fermions and an antifermion). The initial particle has spin $\frac{1}{2}$ and each of the final particles also has spin $\frac{1}{2}$. By now you recognize that the apparent change from angular momentum $\frac{1}{2}$ to angular momentum $\frac{3}{2}$ is not a problem, because of the vector nature of angular momentum. If the neutron's spin is directed "up," for instance, the spins of the final three particles can be up-up-down, for a total of $\frac{1}{2}$ directed up. Other conserved quantities in this decay process are energy, momentum, charge, baryon number, and lepton number.

A final comment about angular momentum: quantum jumps in atoms as well as in particle decay can involve orbital as well as spin angular momentum. Orbital angular momentum doesn't change the even-odd rules, because it is always an integer. In the neutron-decay example just cited, for instance, if the three final particles separate with an orbital angular momentum of 1 unit directed downward, the three spins could all be directed upward. Again the total would be $\frac{1}{2}$ unit of angular momentum directed upward to match the neutron's original spin.

## Charge

The conservation of charge is somewhat simpler to consider than the conservation of energy, momentum, or angular momentum. Checking charge conservation requires merely counting. Charge is simpler than energy because it has neither multiple forms nor a continuous range of values. And it is simpler than momentum or angular momentum because it is a scalar quantity, not a vector quantity (that is, it has magnitude only, not direction). You have already seen several examples illustrating charge conservation, in this and earlier chapters.

Without doubt the most salutary consequence of charge conservation is the stability of the electron. If it were not for charge conservation, an electron could decay into a neutrino and a photon:

$$e^- \not\rightarrow \nu_e + \gamma.$$

(As before, the slash means that this decay does not occur.) The electron cannot decay into lighter charged particle because there are none (at least none that are known). So the electron's only option, if it is to decay at all, is to decay into neutral particles, in violation of charge conservation. In the process displayed above, all conservation laws except that of charge can be satisfied. Searches for electron decay have not found any example. The absence of any observed decay is expressed in terms of a lower-limit lifetime. Present evidence puts this lower limit at $5 \times 10^{26}$ years, much more than a billion times the life of the universe. So the decay of the electron is not something we need to worry about. (Yet how astonishing and wonderful it will be if physicists ever discover that the electron does have some minuscule probability to decay. Such a discovery would send theorists scurrying to their blackboards and computers.)

## Baryon Number

Physicists have reason to think that charge conservation is indeed absolute but that baryon conservation is not. Yet no example of baryon-number change has ever been found. Experiments to date are consistent with the absolute conservation of baryon number. Searches for proton decay (so far in vain) put the lower-limit lifetime of the proton at $10^{25}$ years, nearly as great as the lower limit for the electron. Since the proton is the least massive known baryon, its decay would necessarily violate baryon conservation. It's nondecay is the strongest test we have of baryon conservation. Experimenters continue to search for proton decay—which, if found, would be extraordinarily interesting but less shocking than if the electron were found to be unstable.

## Lepton Number

First, a reminder about some terminology. There are six *leptons*: the electron and its neutrino, the muon and its neutrino, and the tau and its neutrino. Each of these two-particle groups is said to have a *flavor*—the electron flavor, the muon flavor, and the tau flavor, respectively. Since all six of the leptons have spin $\frac{1}{2}$, they are all *fermions*.

For many years, physicists thought that each lepton flavor was separately conserved. (The conservation of flavor means that the number of particles of a given flavor minus the number of antiparticles of that flavor is constant.) This supposition is consistent with the observed decay of the muon,

$$\mu^- \to \bar{\nu}_\mu + e^- + \bar{\nu}_e,$$

in which one particle of the muon flavor (the muon itself) disappears, to be replaced by another muon-flavored particle (the muon neutrino); and in which a particle and an antiparticle of the electron flavor are created. Flavor conservation is also supported by the absence of the decay of a muon into an electron and a photon:

$$\mu^- \not\to e^- + \gamma.$$

This is an event that has never been seen. Experiment puts its probability at less than 1 in $10^{11}$ (one hundred billion) muon decays. You can see that if it did occur, both muon flavor and electron flavor would change.

In fact, to this day no example of lepton flavor change involving charged leptons has been observed. Yet the observation of neutrino oscillation, as described in Chapter 3, shows that one lepton flavor can turn into another, invalidating lepton flavor as an absolute conservation law. The constancy of *total* lepton number (all flavors combined) is nevertheless holding up as an absolute conservation law. In the neutrino-oscillation phenomenon, for example, a neutrino of one flavor turns into a neutrino of another flavor, but there is no change in the number of neutrinos.

Physicists' attitude toward lepton-number conservation is pretty much the same as toward baryon-number conservation. Experiment is consistent with both of these being absolute conservation laws, but it would be no surprise if one or both proved to be only partial, with some small chance (undoubtedly *extremely* small) of violation.

## Color

Color is an important but shy attribute of quarks and gluons that is believed to be absolutely conserved. Leptons and bosons are colorless. So is every composite particle observed in the lab—protons, neutrons, pions, lambda particles, and so on. So color completely escapes direct observation. The evidence for its conservation is indirect, based on the success of the quark theory of strong interactions.

Color, like electric charge, is a quantized attribute of particles that is passed along from one particle to another like a baton in a relay race. It is a bit more complicated than charge because there are three colors (red, blue, and green, by arbitrary convention) and three anticolors, whereas charge is just positive or negative. Inside a proton or neutron is a whirl of colors, as quarks dance back and forth among red, blue, and green, and gluons flit in and out of existence as red-antigreen, blue-antired, and so on. Look at Figure 12 on page 90 to see a possible small part of such a color-laden dance. At every vertex, color is conserved.

## TCP

The final entry in the list of absolutes is a symmetry principle that goes by the name TCP. It's a bit complicated, but, for the sake of completeness, here it is (once again, fasten your seat belt). The principle of TCP invariance says that if you make all three of the following changes of conditions for a physically possible process, the result will be another process that is also physically possible, following the same laws of nature as the original process. The changes or conditions are

> T, *time reversal:* Run the experiment backwards—that is, interchange before and after;
>
> C, *charge conjugation:* Replace all particles in the experiment with their antiparticles and all antiparticles with their particles;
>
> P, *parity, or mirror reversal:* Run the experiment as the mirror image of the original experiment.

Up until the 1950s, physicists believed that the laws of nature were invariant for each of these changes of condition separately. Then came discoveries which shattered that assumption but left the combined principle of TCP invariance intact. Here's an example to clarify its meaning. Let the original process be the decay of a positive pion at rest into a positive muon and a muon neutrino:

$$\pi^+ \rightarrow \mu_L^+ + \nu_{\mu L}.$$

The subscript L is important. It turns out that all neutrinos are "left-handed." This means that if you point your left thumb in the direction the neutrino is moving, the curved fingers of your left hand will show the direction of spin of the neutrino. So L means left-handed spin. Since the muon and the neutrino fly apart in opposite directions and since their total spin must be zero to match the pion's zero spin, the muon, too, must have left-handed spin. Now consider the action of the three changes of conditions. Time reversal (T) will result in a muon and neutrino colliding to produce a pion. Charge conjugation (C) will change the positive muon to its antiparticle, a negative muon; the positive pion to its antiparticle, a negative pion; and the neutrino to an antineutrino. Mirror inversion (P) will change left-handed spins to right-handed spins. The prediction of the invariance principle, then, is that the following process is a physically possible one:

$$\bar{\nu}_{\mu R} + \mu_R^- \rightarrow \pi^-.$$

Figure 25. Neutrinos are left-handed.

There is every reason to believe that this is a possible process, although it is surely not a practical one. To see it would require that beams of negative muons and antineutrinos be fired at each other with equal and opposite momenta and just the right energy. If by some miracle this could be done, a negative pion would occasionally be produced (the chance of this happening can be calculated exactly).

Although this example does not lead to an experimental test of the TCP theorem, it does reveal one thing that has been checked: antineutrinos are all right-handed. The best tests of TCP invariance are its predictions that every antiparticle should have identically the same mass as its companion particle, and that every antiparticle that is unstable should have the same lifetime as its companion particle. These predictions have been tested to a high degree of precision. TCP invariance also has strong theoretical underpinning. If it were shown to be less than absolute, by whatever small margin, the entire structure of quantum theory would be revealed as resting on shaky ground.

## Partial Conservation Laws and Invariance Principles

You might think that a physical quantity is either conserved or it isn't—that a partial conservation law makes as much sense as partial pregnancy. You would be right, in a way, but physicists have learned that some quantities are conserved in processes governed by certain interactions but not in processes governed by other interactions. That's what leads to the idea of partial conservation laws. For instance, quark flavor is conserved in strong and electromagnetic interactions but not in weak interactions. And a quantity called isospin (defined below) is conserved by the strong interactions but by neither the electromagnetic nor the weak interaction.

Similarly, certain invariance principles are respected by some but not all interactions. Charge conjugation invariance, for example (particle-antiparticle interchange), is valid for the strong and electromagnetic interactions but not for the weak.

There is a rule here: the stronger the interaction, the more numerous the constraints. The *strong* interaction is hemmed in by the most conservation laws and the most invariance principles; the *electromagnetic* interaction, by slightly fewer; and the *weak* interaction, by still fewer. Is *gravity*, the weakest of all interactions, an even more flagrant violator of conservation laws and invariance principles? That's a wonderfully interesting question to which we do not yet know the answer, because, so far, gravitational effects have not been detected at the level of particle reactions.

Still in an experimental netherworld, though much on the minds of theorists, is a particle called the *Higgs boson* (named after one of its inventors, the Scottish physicist Peter Higgs). It is the predicted quantum manifestation of a still-hypothetical field believed to permeate all space and to account, through its interactions, for particle masses. That's a big job to place on its shoulders. In addition, it may have something to do with the exceedingly weak violation of time-reversal invariance that has been detected and for an also decidedly weak violation of lepton flavor conservation. No endeavor in particle physics is more exciting than the ongoing search for the Higgs. If this particle is found and revealed to do the jobs assigned to it, it will be one more example of a deeper simplicity coming along to slightly undermine an appealing but less deep simplicity. (In this case, the fundamental Higgs field permeating all of space would account for the slight imperfection of time-reversal invariance.)

## Quark Flavor

Quarks are more flavorful than leptons. Whereas the six leptons are divided into three groups of two, each of the groups being assigned a flavor (electron, muon, and tau), every one of the six quarks has to be assigned its own flavor. So, in not very felicitous language, we have for the quark flavors upness, downness, charm, strangeness, topness, and bottomness. These quark flavors are conserved in strong interactions (as well as in electromagnetic interactions). Consider, for example, the collision of one proton with another that produces a proton, a neutron, and a positive pion,

$$p + p \rightarrow p + n + \pi^+.$$

Here, only up quarks and down quarks are involved. In terms of quark constituents, this reaction can be written

$$uud + uud \rightarrow uud + udd + u\bar{d}.$$

Initially, there are four up quarks and two down quarks. Finally, there are four up quarks, three down quarks, and one down antiquark, so the flavors balance (the antiquark must be counted as having negative flavor). Here's one more example, a proton-neutron collision that produces strange particles:

$$p + n \rightarrow n + \Lambda^\circ + K^+.$$

In terms of quark constituents, this reaction lines up as follows (s stands for the strange quark):

$$uud + udd \rightarrow udd + uds + u\bar{s}.$$

Three up quarks and three down quarks result in three up quarks, three down quarks, one strange quark, and one strange antiquark. Again, the flavors balance.

These two examples involve the strong interactions. We know the strong interactions are at work because the reactions occur with high probability. Consider, however, the decay of a lambda particle, which is a slow process governed by the weak interaction:

$$\Lambda^\circ \rightarrow p + \pi^-.$$

In terms of quark constituents, this decay process reads

$$uds \rightarrow uud + \bar{u}d.$$

What has happened here is that the strange quark has disappeared, to be replaced by a down quark. So the weak interaction violates the law of quark flavor conservation—which is, accordingly, a partial conservation law.

### Isospin

The concept of isospin goes back to the 1930s. Soon after the discovery of the neutron in 1932, it became evident that this new particle and the proton had a great deal in common, despite their difference in charge. They had nearly the same mass and seemed to have the same strong interaction. The proton and neutron came to be regarded as two states of a single underlying particle, the nucleon, to which the mathematics of a spin-one-half particle with its two orientations of spin could be applied (thus the name *isospin*, which otherwise has nothing to do with spin). Later other particle "multiplets" were found, such as the pion triplet and the xi-particle doublet (and some singlets such as the lambda particle). The law of isospin conservation states that a total isospin can be assigned to any group of particles and that this quantity remains unchanged when the particles interact strongly, but can change if the particles interact via the electromagnetic or weak interactions. The idea is a little easier to understand if restated as an invariance principle. In the language of invariance, it says that if one member of a multiplet is replaced by another member of the same multiplet, the strong interactions remain unchanged. This predicts, for example, that the proton and neutron have identical strong interactions, as do the positive, negative, and neutral pions—conclusions that could also be reached from the law of quark flavor conservation. I discuss isospin as a separate idea partly because of its history, and partly because it deals with the particles that are actually observed, not with the unobserved quarks. Isospin invariance is clearly violated by the electromagnetic interaction because particles within a given multiplet differ in charge (and also differ slightly in mass).

## P and C

Until the mid-1950s, physicists—with an insufficient supply of their usual caution—took it as self-evident that each of the three invariance principles T, C, and P was an absolutely valid principle. The principles were tested for strong and electromagnetic interactions. They were such lovely principles that they had to be valid for the weak interactions as well—or so it seemed. In 1956, physicists collectively turned pink with embarrassment when two young Chinese-American theorists, Tsung-Dao Lee (then twenty-nine and at Columbia University) and Chen Ning Yang (then thirty-three and at the Institute for Advanced Study in Princeton), pointed out that there was no experimental evidence whatsoever for the validity of parity conservation in *weak* interactions. They suggested that P violation would help clear up an oddity that had appeared in particle data,* and called on experimenters to test the validity of the principle. That same year Lee's Columbia colleague Chien-Shiung Wu† undertook an experiment whose results the next year provided dramatic evidence that P conservation, although "self-evident," is not true. Almost at once, other groups using other methods confirmed her finding.

Parity conservation, or space-inversion invariance, can be stated

---

* It seemed that two particles alike in mass and other properties had different parity. These "particles" we now know to be a single particle, the kaon. In 1955–56, while in Germany on a year's leave of absence from Indiana University, I was one among many theorists around the world puzzling over this oddity. Like almost all the rest, I didn't think to ask whether parity might not be conserved.

† Chen Ning Yang is known to all his colleagues as "Frank." Tsung-Dao Lee is widely called "T.D." Chien-Shiung Wu (who died in 1997 at age eighty-four) was known to most as "Madame Wu" (though called "Chien-Shiung" by her close associates).

this way: the mirror view of a possible process is a possible process. If you look at this page in a mirror, you will see backward writing—not normal, to be sure, but not impossible. Backward type could easily be designed and set to produce backward writing (which, in a mirror, would look normal). Whoever paints

# AMBULANCE

on an ambulance is doing just that. Space-inversion invariance holds true in the everyday world. What you see in a mirror (a flat mirror) may be unlike anything you have seen before; it may be quite strange; but it is not impossible—it violates no physical laws. Another way to understand space inversion is to imagine watching a film in which each frame has been reversed left to right. Watching the film, you wouldn't have much trouble knowing that the film is inverted. You might notice that nine out of ten actors were left-handed, or that the men buttoned their shirts from right to left, or that cars in the United States drove on the right, or that signs appeared with backward writing. Yet you would have to conclude that nothing in the inverted scenes you were watching was obviously impossible.

By contrast, a mirror view of the experiment carried out by Madame Wu is a view of an impossible process. She and her team aligned spinning cobalt 60 nuclei* in a magnetic field at extremely low temperature. An aligned nucleus is shown on the left side of Figure 26. If the curved fingers of your right hand follow the arrow that designates the rotation of the nucleus, your right thumb points upward in the direction defined as the axis of spin. You can say that the "north pole" of the nucleus is on top. What Madame Wu and her group observed was that most of the electrons emitted in the beta decay of this nucleus (now imagine many such nuclei) shoot downward, in the direction of the "south pole." The right side of Figure 26 shows a mirror view of the process: the north pole of the nucleus is on the bottom (again let your right hand define the di-

---

* Cobalt 60 is an isotope widely used in medical applications. In the aftermath of a nuclear explosion in wartime, it could constitute a threat to health.

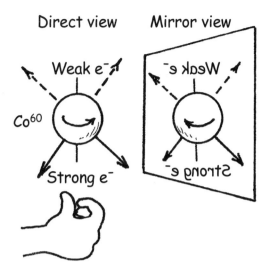

Direct view    Mirror view

Weak e⁻

Co⁶⁰

Strong e⁻

Figure 26. Beta decay of cobalt 60.

rection of the axis of spin), and the electrons are shooting out mainly in the north-pole direction. But that is inconsistent with what was observed in the laboratory. The mirror view is a view of the impossible. Parity, in this weak-interaction process, is not conserved.

The experiment seems so simple and so direct. Why hadn't it been done earlier? Partly because no one thought it needed doing. And partly because it really was harder than my discussion suggests. If the magnetic field were not strong and if the temperature were not extremely low (it was about one-hundredth of a degree above absolute zero), the nuclei would not remain aligned.* They would flop around, with as many having their spin up as down. Then electrons would be observed to be ejected equally up and down, even though each nucleus is ejecting its single electron in one direction relative to its spin.

On January 4, 1957, a group of Columbia physicists gathered for their regular weekly luncheon at a Chinese restaurant near the university. T. D. Lee reported to the group what he had recently learned from Madame Wu: that electrons were shooting out asymmetrically from the

---

* Wu and her team conducted their experiment not at Columbia University but at the National Bureau of Standards in Washington, where the needed low temperatures could be achieved.

aligned cobalt 60 nuclei in her experiment. Leon Lederman (later to win a Nobel Prize for other work) realized that another experiment that Lee and Yang had suggested to check on parity conservation could probably be carried out at the Columbia cyclotron. That evening he and his graduate student Marcel Weinrich went up to the cyclotron lab north of New York City to get the experiment started. They phoned a young faculty colleague, Richard Garwin (who had missed the luncheon because he was traveling), and Garwin joined them the next day. A mere three days later they had discovered that the muon's neutrino is single-handed (either left-handed *or* right-handed) and that the neutrino's weak interaction violates both space-inversion and charge-conjugation invariances.*

Lederman and his colleagues used the Columbia cyclotron to create positive pions, then studied the decay of these pions into muons and neutrinos, and the subsequent decay of the muons into electrons and more neutrinos. The decay of a positive pion is shown in the center of Figure 27. The shaded circle shows where the pion was before it decayed. Shooting upward from that spot is a positive muon, which the experimenters showed to be (within experimental uncertainty) always single-handed.† Let the curved fingers of your left hand follow the arrow that designates the muon's rotation. Then your left thumb points in the direction of the muon's travel. Now enter two well-established conservation laws, those of momentum and angular momentum. Because of momentum conservation, the unseen neutrino is known to be flying downward, opposite to the direction of the muon's flight. And because of angular momentum conservation (taking account of the fact that the original pion is spinless), the neutrino's spin must be opposite to the muon's spin. Therefore, the neutrino has the same handedness as the muon. (Check this by using your left hand again to follow the neutrino's

---

* At the time of this experiment, the muon neutrino and electron neutrino were not yet known to be distinct. It was for discovering the muon neutrino's distinctness five years later that Lederman shared a Nobel Prize.

† The Columbia team determined the alignment of the muon's spin by measuring properties of its subsequent decay. I won't pursue that aspect of their experiment here.

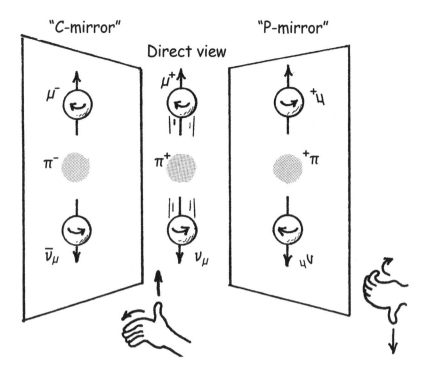

Figure 27. Positive pion decay with "C-mirror" and "P-mirror" views.

spin and direction of flight.) If parity were conserved, half of the neutrinos created would be left-handed and half would be right-handed. Experiment indicated that they all were single-handed (meaning that parity conservation is as totally violated as it could be!).

Shown on the right side of Figure 27 is a regular mirror, or "P-mirror." In the mirror you see a right-handed neutrino flying downward. This is a view of the impossible, since neutrinos are left-handed. So, no P-invariance. On the left side of the figure is a "C-mirror," showing the result of applying charge conjugation (particle-antiparticle inversion) to the process. In the C-mirror, a negative pion (the antiparticle of a positive pion) decays into a negative muon (the antiparticle of a positive muon) and a left-handed antineutrino. But if neutrinos are left-handed, antineutrinos must be right-handed. The C-mirror, too, is showing an impossible process. So, no C-invariance.

To reiterate a point: the maximal violation of P-invariance and C-invariance revealed in these experiments applies only to the weak interaction, which controls charged-pion decay, muon decay, and beta decay. P-invariance and C-invariance were, and still remain, valid principles for electromagnetic and strong interactions.

### T and PC

Shown in Figure 28 is a "CP-mirror," which interchanges left and right and *also* interchanges particles and antiparticles. The decay of a positive pion into a positive muon and a left-handed neutrino,

$$\pi^+ \rightarrow \mu^+ + \nu_{\mu L},$$

when seen through the CP-mirror, becomes

$$\pi^- \rightarrow \mu^- + \bar{\nu}_{\mu R},$$

which is the decay of a negative pion into a negative muon and a right-handed antineutrino. But that is exactly what is observed for negative pion decay. This makes it appear that combined PC invariance is valid for the weak interaction, and therefore for all interactions. If it is, then time-reversal invariance, or T-invariance, should also be absolutely valid because of the strong evidence that TCP, the combination of all three invariances, is indeed absolutely valid.

After the startling revelations of 1957, physicists got used to the idea that although P-invariance and C-invariance are separately violated by the weak interaction, the combination PC, along with T and TCP, are absolutely valid. This faith in a revised simplicity lasted only until 1964, when two Princeton physicists, Val Fitch and James Cronin, working at the Brookhaven accelerator on Long Island, shook up physics once again. They found that a long-lived version of the neutral kaon (having a lifetime of $5 \times 10^{-8}$ second), which decays readily into three pions, decays occasionally—about one time in five hundred—into two pions. Theory showed that this could happen only if CP-invariance is violated. This was another startling conclusion, which implied in turn that time-reversal invariance is likewise not an absolutely valid principle.

The violation discovered by Fitch and Cronin (which earned them

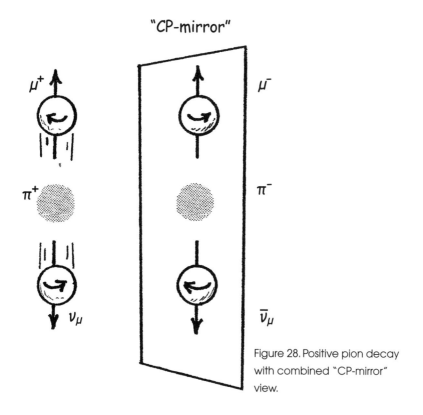

"CP-mirror"

Figure 28. Positive pion decay
with combined "CP-mirror"
view.

the 1980 Nobel Prize) is "weaker than weak"—although it later showed
up more strongly in the decay of the B particle, a very heavy meson
(mass 5,279 MeV) that contains a bottom (or antibottom) quark. The
implications of CP violation are quite startling. Val Fitch likes to say
that it is the "reason we are here." It was a discovery more unsettling to
physicists even than the earlier discoveries of the violations of C and P
invariance separately. This is because it implies a basic difference between matter and antimatter. Worlds made of matter and antimatter
don't behave quite identically. For that reason, observations on Earth
could—in principle—reveal whether a distant galaxy is made of matter
or antimatter. There is no evidence to date for antimatter galaxies, and
pretty good evidence that our universe is made entirely of ordinary matter. According to present theory, there were, shortly after the Big Bang,
a nearly, but not exactly, equal number of protons and antiprotons.

These particles, in a hot, dense soup, largely annihilated each other as they repeatedly collided. What was left over to form the universe we live in was the slight excess of matter over antimatter, no more than one part in a billion. If it were not for CP violation, so goes the theory, there would have been an exactly equal number of baryons and antibaryons (or, equivalently, of quarks and antiquarks). All of them would have been annihilated and the remaining universe would have consisted of photons and perhaps neutrinos. In such a universe, there would be no galaxies, no stars, and no us.

## Symmetry

"Symmetry" is a term most of us use regularly, often to mean "balance," or perhaps "harmony," or sometimes even "beauty." A mathematician or physicist sees these same attributes in symmetry, but defines the idea this way: something has symmetry if one or more of its properties remains unchanged when some other property is changed.

A long straight railroad track has a kind of *translational symmetry*. It

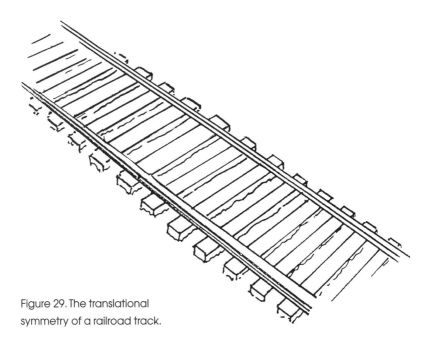

Figure 29. The translational symmetry of a railroad track.

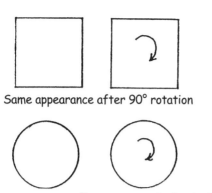

Same appearance after 90° rotation

Same appearance after any amount of rotation

Space-inversion symmetry

Figure 30. Types of symmetry.

remains unchanged if you displace it along its length by the distance separating one tie from the next (or change your viewpoint in the same way). A square has a kind of *rotational symmetry*. It is restored to its original form if you rotate it through 90 degrees (or any multiple of 90 degrees). A circle has a more encompassing kind of rotational symmetry. It remains unchanged if rotated through any angle at all. This symmetry no doubt played a role in the thinking of Aristotle and other ancients, to whom the circle was the most perfect of plane figures. A mask that is perfectly balanced between its left and right sides has *space-inversion symmetry*. You can't tell whether you are looking at the mask or its mirror image, for they are identical.

Imagine closing your eyes while a friend makes a change—or *perhaps* makes a change—and then opening your eyes. If you were looking at a square and it looks just the same after you open your eyes, you can't tell whether your friend rotated the square or not. If you were looking at a photograph of a perfectly symmetric face and it looks the same after you open your eyes, you can't tell whether your friend substituted a mirror view of the face for the original photo. If you were in a caboose looking back along the track, and it all looked just the same after you opened your eyes, you couldn't tell whether your friend had arranged for the train to move forward a bit.

Now suppose you and your friend are in the depths of empty space when you close your eyes. When you open them, everything seems exactly the same. Are you where you were before, or have you moved? You can't tell. This illustrates a symmetry called the homogeneity of space (for which we have good evidence). Space is everywhere the same. There is no preferred location. But what does it mean to say that everything "seems" the same? You have brought with you in your spacecraft lots of laboratory equipment and you proceed to do experiments to see if any of them come out differently than they did previously. No, all results are the same. Your friend now reveals that you have moved. You conclude that the laws of nature are the same in one place as another. Your efforts have yielded an important link:

invariance ⟷ symmetry.

Related to the *symmetry* that we call the homogeneity of space is the *invariance* of the laws of nature. Physicists believe that this illustrates a universal link—that every invariance principle is linked to a symmetry principle. In most cases, the link is easy to see. If, for instance, the mirror world has properties identical to those of the real world (space-inversion symmetry), then the laws of nature have space-inversion invariance. (We know that in fact only *some* laws have this invariance.) In many cases, symmetry and invariance are almost identical ideas. This is true whenever it is the laws of nature that are the unchanging "property" of the symmetry.

A subtler link is

symmetry ⟷ conservation.

In 1915, Emmy Noether (pronounced roughly as "NUR-ter," but with the first r suppressed), a guest researcher and lecturer at the University of Göttingen, proved a remarkable theorem that has been a pillar of

mathematical physics ever since.* She showed that for every continuous symmetry, there is an associated conservation law. (A continuous
symmetry is one that involves a change of conditions of any magnitude
whatsoever, unlike the "discrete symmetries" of the railroad track, the
rotating square, or the symmetrical face.) Symmetry becomes the "glue"
that links invariance principles and conservation laws:

invariance $\longleftrightarrow$ symmetry $\longleftrightarrow$ conservation.

To illustrate the idea, let me try to explain how the conservation of momentum is related to the homogeneity of space (which is a continuous
symmetry).

I start with a book on a table. Any book will do, but the table must
be special: perfectly flat and perfectly frictionless. Then I can forget all
about gravity, for it can have no effect on the horizontal motion of the
book. I place the book carefully on the table so that it is motionless. It
remains motionless. That seems reasonable, even "obvious," until I raise
this question: Given that the book contains trillions of atoms, each
one tugging on others, some in one direction, some in another direction, some strongly, some weakly, how does it happen that all of these
interactions among the book's atoms miraculously have no effect? Isaac
Newton answered this question more than three hundred years ago with
a law of compulsion that we now call Newton's third law. It states that
all forces in nature come in balanced pairs. You may have heard it
phrased this way: for every action there is an equal and opposite reaction. If atom A exerts a force on atom B, then atom B exerts an equal
and opposite force on atom A. All such pairs of forces add (as vectors)
to zero. Atom A moves. Atom B moves. But the book—the collection
of all the atoms—doesn't move.

A modern approach to the "problem" of the quiescent book is to

---

* The great Göttingen mathematician David Hilbert had to joust with his university administration for four years before he succeeded in getting Emmy Noether ap-
pointed to a proper academic position. Because she was Jewish, she lost that position in 1933. She then moved to Bryn Mawr College in Pennsylvania, where she
was a professor until her death in 1935.

look for a symmetry principle, some deeper reason that will, in effect, explain *why* Newton's third law is true. The symmetry is the homogeneity of space. The related invariance principle states that the laws of nature are the same in all locations. This leads to the conclusion that no aspect of the motion of an isolated object depends on where the object is. The book on the table is isolated in the sense that no outside effect is pushing it horizontally in any direction. Suppose that the book, after being at rest at point X, "decides" to start moving. Later it will pass through point Y. Its state of motion at Y, where it is moving, is different from its state of motion at X, where it was at rest. This is inconsistent with the original symmetry principle. If the book starts moving, then its state of motion depends on where it is. So, frustrated in its "desire" to move, it stays put.

This discussion can be replaced by sound mathematics, which shows that the aspect of motion that cannot change spontaneously if space is homogeneous is momentum. If the book, free of outside influences, is stationary at one place and then moving at another place, it has changed its momentum. Its momentum is not conserved. Momentum conservation *prohibits* its motion if it is initially at rest. If, on the other hand, it is set in motion when first placed on the table, it will continue moving with the same momentum. Then, again, no aspect of its motion depends on where it is.

It is a stunning idea that the bland sameness of empty space accounts for the conservation of momentum, but it does. It is the featurelessness of space that makes Newton's third law a necessity. An indication of the greater depth of the symmetry approach is that momentum conservation, a valid principle of classical physics, remains valid in the modern theories of relativity and quantum mechanics, whereas Newton's third law, expressed in terms of action and reaction forces, needs change. It can be applied directly only in classical physics, not in the modern theories.

The conservation of angular momentum is related to another symmetry of space: its *isotropy*, the fact that there are no preferred directions in space. If a compass needle in remote space, free of magnetic fields and other outside influences, were to start rotating spontaneously, its angular momentum would change and its motion would imply that spatial directions were not all equivalent. Even more remarkably, energy conservation rests on the symmetry of time and the invariance of the laws of physics to change of time. So all three conservation laws concerned with motion—momentum, angular momentum, and energy—rest simply on the fact that spacetime is completely uniform, the same everywhere.*

Physicists now believe that *every* conservation law rests ultimately on a symmetry principle. They have demonstrated this for charge conservation, whose related symmetry is the sameness of physical results when certain unobservable changes are made in the wave functions of charged particles. For some of the particle conservation laws, however, the underlying symmetry remains to be discovered.

One may reasonably claim that conservation laws, being based on the properties of empty space and on other symmetries, are the most profound expressions of physical law. On the other hand, they may be, as once claimed by the eminent mathematician and philosopher Bertrand Russell,† mere "truisms," because, he asserted, the conserved quantities are *defined* in just such a way that they must be conserved. I like to think that both points of view can be correct. If the aim of science is the self-consistent description of nature using the simplest set of basic assumptions, what could be more satisfying than to have basic assumptions so elementary, even "obvious" (such as the uniformity of space and time), that the laws derived from them can be called truisms? The scientist, inclined to call most profound that which is simplest and most general, is not above calling a truism profound. And is it not true that the discovery of *anything* that remains constant throughout all processes of change is a remarkable achievement, regardless of the arbitrariness of definition involved?

Still an open question is whether conservation laws are all that are needed to describe nature, whether everything that happens—all the laws of change, all the laws of compulsion—will be shown to rest on conservation laws and, ultimately, on symmetry.

---

* The theory of relativity has joined these three laws into a single conservation law in the four-dimensional world.

† Bertrand Russell, *The ABC of Relativity* (New York: New American Library, 1959).

chapter 9

# Waves and Particles

$A$ s a college student in France before the First World War, Prince Louis-Victor de Broglie (pronounced, roughly, "Broy") concentrated first on history, with the idea of entering the diplomatic corps. Then he fell in love with theoretical physics, abandoned the history research project he had been assigned, and earned his undergraduate degree in physics in 1913 at age twenty. That was the year in which Niels Bohr published his quantum theory of the hydrogen atom. Later, in his Nobel lecture of 1929, De Broglie would speak of his attraction to "the strange concept of the quantum, [which] continued to encroach on the whole domain of physics."

After serving in the army, de Broglie began his graduate studies at the University of Paris, and, in 1924, submitted a doctoral dissertation that contained a revolutionary idea: in the quantum world, he said, waves are particles and particles are waves. It was an idea with staying power. This *wave-particle duality* remains central in quantum physics. De Broglie said afterward that two lines of thinking led him to this idea. First was the growing realization among scientists that X rays exhibited both wave and particle properties. Until Arthur Compton's 1923 work on the scattering of X rays by electrons in atoms, most physicists were reluctant to accept the reality of Einstein's photon. (Even in 1924 the photon was, to Satyendra Bose, still a hypothesis.) There was no doubt that X rays were electromagnetic waves. They exhibited standard wave properties such as diffraction and interference. Compton's work—as well as a phenomenon called the *photoelectric effect*, in which individual photons eject electrons from a metal surface—made clear that electromagnetic waves also have "corpuscular" properties (to use the terminology of the time). If waves (such as X rays) can show particle properties, mused de Broglie, why can't particles (such as electrons) show wave properties?

De Broglie also took note of the fact that in the classical world, waves, but not particles, are quantized. By this he meant that piano and violin strings, the air in organ pipes, and many other systems involving waves vibrate at selected frequencies, not arbitrary frequencies. No such quantization was known for particles in the classical world. This made him wonder whether the observed quantization of energy levels in atoms might be the result of vibrating "matter waves"—whether, in effect, the atom is like a musical instrument.

So de Broglie was led to suggest that electrons—and, by inference, other particles—have wavelike properties such as frequency and wavelength. Three years later, in 1927, Clinton Davisson and Lester Germer, working at Bell Labs in the United States, and, independently, George Thomson (son of J. J. Thomson, the electron's discoverer), working at Aberdeen University in Scotland, established the wave nature of electrons by observing that electron beams striking crystalline solids exhibit diffraction and interference effects (See Figure 31). From the observed patterns, they could measure the wavelength of the electrons. But even before that confirmation, theorists had accepted the wave idea. In 1926, for instance, the Austrian physicist Erwin Schrödinger had proposed a wave equation which, when applied to the hydrogen atom, yielded the observed quantized energies, validating de Broglie's supposition about the "reason" for quantized energy.*

---

* The wave-particle duality led to a slew of Nobel Prizes: Einstein in 1921 for the photoelectric effect, Compton in 1927 for the "Compton effect" (photon scattering from electrons), de Broglie in 1929 for discovering the wave nature of electrons, Schrödinger in 1933 for "new forms of atomic theory" (his wave equation), Davisson and Thomson in 1937 for the diffraction of electrons by crystals, and Max Born in 1954 for relating the wave function to probability.

Clinton Davisson (1881–1959), left, and Lester Germer (1896-1971) with a tube used in their electron-diffraction work, 1927. Photo courtesy of Lucent Technologies' Bell Laboratories and AIP Emilio Segrè Visual Archives.

It is not hard to see why the wave-particle duality was at first a very unsettling idea for physicists. Waves and particles seem to be very different concepts with little in common. Nearly everyone is familiar with both particles and waves in the everyday world. For most purposes, a baseball, tennis ball, golf ball, or speck of dust (or an asteroid coursing through space) is a particle. It is small (relative to dimensions of interest around it); it has mass; it has a particular location at any time; and it may have momentum and energy. Waves in the world around you include water waves, sound waves, radio waves, and light waves. A water wave exhibits the essential features of a wave. It is *not* small, for it is spread out. You can't pinpoint its location at a particular spot (although

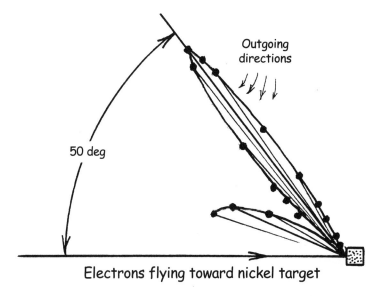

**Electrons flying toward nickel target**

Figure 31. Experimental results of Davisson and Germer showing that electrons of 54 eV, after striking a nickel crystal, emerge mostly in a certain direction, because of diffraction and interference of the electron waves. Image adapted from *Nobel Lectures, Physics* (Amsterdam: Elsevier, 1965).

it can be localized in a region). It oscillates. It has a wavelength, which is its crest-to-crest or trough-to-trough distance;* a frequency, which is the number of cycles of oscillation it completes in a unit of time; and an amplitude, which is the strength of its vibration. It may be a traveling wave, in which case it has a speed; or it may be a standing wave, a vibration that doesn't propagate anywhere, like the vibration of a guitar string or the air in a flute (or the sloshing of water back and forth in a

---

* A water wave is called a *transverse* wave because the water moves mainly up and down transverse to the direction of motion of the wave. Radio waves and light waves are also transverse. A sound wave is a longitudinal wave. This means that the vibrating material is moving back and forth parallel to the direction of wave motion. A longitudinal wave still has crests (locations of above-average density) and troughs (locations of below-average density), so it has a well-defined wavelength, as well as frequency and speed.

Figure 32. A "localized" pulse running along a rope.

bathtub). If it travels, it does so at a speed that doesn't depend on how much energy it is carrying.*

So particles (in the classical macroworld) are localized and waves are not. Waves have wavelength and frequency and amplitude, and particles (apparently) do not. A particle, when it travels, goes from one place to another. A wave, although it may propagate, does so without dragging material along. (The water in a water wave or the air in a sound wave vibrates about a fixed location while the wave moves on.) And particles move faster if they have more energy, whereas waves (of a given kind) propagate at the same speed regardless of their energy. Yet even classical waves and particles *do* have some things in common. They can transmit energy and momentum from one place to another. And waves can be at least partially localized. Consider a single pulse running along a rope that has been snapped at one end, or the reverberating tone of an organ pipe within a cathedral.

For waves and particles to be joined in the subatomic world, both concepts had to give a little. Particles can no longer be exactly localized—that's the biggest change of all. And waves became somewhat more "material." The thing doing the vibrating is a "field," an entity that possesses energy and momentum, unlike the immaterial "ether" referred to in the nineteenth century and before. The wave can be partially localized to mimic, in some degree, a particle.

---

* The rule that wave speed doesn't depend on wave energy is precisely true for light waves and true to very good approximation for most other waves. Shock waves provide an exception to the rule. They are more energetic and move faster than ordinary sound waves.

## The De Broglie Equation

In his dissertation, de Broglie introduced an equation of startling simplicity that proved to be as momentous as Einstein's $E = mc^2$. It is written

$$\lambda = \frac{h}{p}.$$

On the left is wavelength $\lambda$ (lambda). On the right is Planck's constant, $h$, divided by momentum, $p$. (In classical theory, momentum is mass times velocity, $p = mv$, so an object has more momentum if it is more massive or if it is moving faster. According to relativity theory, a particle can have momentum even if it has no mass.)* The de Broglie equation links, across the "equals" sign, a wave property (wavelength) and a particle property (momentum). It joins concepts that classically seem to have little to do with each other. And the joining glue is Planck's constant. This means that the link is a *quantum* link.

Any equation, this one included, does more than summarize ideas. It provides a recipe for calculation. Given the measured wavelength of an electron beam, for instance, a physicist can calculate the momentum of the electrons in the beam. Or a researcher who knows the momentum of neutrons in a beam can calculate the wavelength of the neutrons.

The appearance of momentum "downstairs" (that is, in the denominator) on the right side of the de Broglie equation is significant. It means that as momentum gets larger, wavelength gets smaller. This is part of the reason that modern accelerators are so large and so expensive. Researchers want to use very small wavelengths in order to probe the tiniest subatomic distances, so they must accelerate particles to great energy and huge momentum.

According to Bohr's first theory of the hydrogen atom, an electron in the lowest-energy state (the ground state) in the atom moves with a speed of about 2 million ($2 \times 10^6$) meters per second. One can multiply

---

* The relativistic definition of momentum for a massless particle is $p = E/c$, where $E$ is the particle's energy and $c$ is the speed of light.

this speed by the electron's mass to get its momentum, then use de Broglie's equation to calculate the electron's wavelength. The answer is about $3 \times 10^{-10}$ m. This result excited de Broglie because this calculated wavelength is the same as the calculated circumference of the circular orbit of the lowest-energy state in the hydrogen atom, according to Bohr's theory. This suggested to de Broglie a principle of *self-reinforcement,* the idea that the electron's wave executes an integral number of cycles of oscillation as it circles once around the orbit, so that a crest in the wave meets and reinforces a crest after one trip around. Suddenly it appeared that the wave nature of the electron could explain why that state of motion had the size and energy that it did. Any other wavelength, went the reasoning, and a crest would not meet a crest. Indeed, after many trips around the orbit, the wave would wipe itself out (as shown on the left side of Figure 33).

As it turned out, de Broglie's idea of a wave traveling around an orbit was too simplistic. Quantum mechanics, once fully developed, showed that the electron wave in the atom is three-dimensional, spread out over space, not just stretched along an orbit, and that the electron itself must therefore be viewed as spread out, not executing a specific orbit. Nevertheless, de Broglie's calculated wavelength correctly gave the approximate size of the hydrogen atom and opened the door to understanding the atom in terms of spread-out waves.

What about people? Do we have wavelengths, too? Yes, but far too small to measure. The "fuzziness" of our size introduced by our wave nature is truly minuscule. A person strolling at a speed of 1 m/s has a wavelength of about $10^{-35}$ m, many, many orders of magnitude smaller than the size of a single atomic nucleus. Why so tiny? Because a person has such enormous momentum (compared with an electron). True, the person's speed is small, but the mass, relative to the mass of a single particle, is astronomical. What about something smaller and slower? It doesn't help much to go to a smaller, slower creature. A one-gram bug crawling along at one meter per year has a wavelength of around $10^{-23}$ m, still imperceptibly small. So people and baseballs, and even bacteria, have sharply defined edges and no discernible wave effects. Yet it is the wave nature of the electron that gives bulk to every atom and prevents people, baseballs, and bacteria from collapsing.

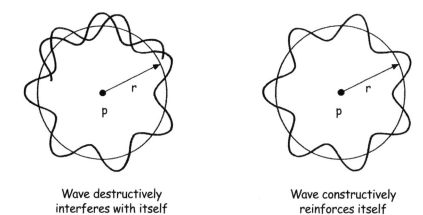

Wave destructively
interferes with itself

Wave constructively
reinforces itself

Figure 33. De Broglie's idea of a wave interfering with itself.

To learn an equation and to appreciate and understand it are two different things. Anyone can commit to memory in a few moments the equations $E = mc^2$ and $\lambda = h/p$, but what do these two equations really mean? Why are they so significant? As an aid to learning to "feel" their meaning, let's compare them. Einstein's mass-energy equation is one of the fundamental equations of relativity; de Broglie's wave-particle equation is a fundamental equation of quantum theory. The first contains the universal constant $c$, the speed of light, which is, in a sense, the fundamental constant of relativity theory: $c$ provides the link between space and time and it sets the "scale" of the theory. Anything moving at much less than the speed of light is described by classical physics; anything moving near or at the speed of light shows the new effects of relativity.

The second equation contains the universal constant $h$, Planck's constant, which is the fundamental constant of quantum theory, and indeed appears in every quantum equation: $h/2\pi$ is the unit in which angular momentum is measured, and $h$ provides the "scale" of the subatomic world—it determines the sizes of atoms and the wavelengths of particles.

In Einstein's equation, $E$ and $m$ are called variables, for unlike $c$ they can take different values for different particles. Similarly, $\lambda$ and $p$ are the variables of de Broglie's equation. Most important for providing new insights into nature, both equations perform a synthesis. Mass and energy, which prior to Einstein's work were thought to be distinct and unrelated concepts, are drawn together into a simple relation of proportionality by Einstein's equation. De Broglie's equation provides an analogous synthesis of wavelength and momentum, ideas believed to be distinct and unrelated prior to de Broglie's work (despite hints from the hypothesized photon).

The positions of the variables in the equations are of the utmost significance. The appearance of $m$ upstairs in Einstein's equation means that more mass has more energy, or, conversely, the creation of more mass requires more energy. As I mentioned earlier, the fact that $p$ is downstairs in de Broglie's equation means that a particle with more momentum has a shorter associated wavelength. The lighter a particle and the more slowly it moves, the greater its wavelength; therefore, the more evident its wavelike properties will be.

Finally, the magnitudes of the constants in the two equations have great import: $c$ is large and $h$ is small, relative to the "normal" magnitudes encountered in the everyday macroscopic world. Thinking in self-centered human terms, then, a little mass corresponds to a great deal of energy, because $m$ is multiplied by the big number $c^2$. For most energy changes in the everyday world, mass changes are too small to notice. (One gram of mass converted to energy was enough to devastate Hiroshima in 1945.)

And because of the smallness of $h$, quantum wave effects don't impinge on our senses. The wavelength associated with a normal momentum is infinitesimal. The size of these constants is related to the late appearance on the human scene of the theories of relativity and quantum mechanics. Precisely *because* the fundamental constants of these theories are so far removed from ordinary human experience, there was little chance that the theories would have been formulated before experimental techniques available to the scientist had extended the range of observation far beyond the range of direct human perception.

There is, by now, a great body of evidence for the wave nature of particles and for the particle nature of radiation. The earliest such evidence was provided by the absorption of electromagnetic radiation only in quantum lumps (the so-called photoelectric effect) and by the scattering of X rays by atomic electrons, both of which phenomena supported the photon idea. The photon is, in a way, the "perfect" exemplar of wave-particle duality, for it clearly has wavelength and frequency (it *is* light) and it clearly is born and dies (is emitted and absorbed) as a particle. Its dual character is revealed, for example, in the decay of a neutral pion, represented by

$$\pi^\circ \rightarrow \gamma + \gamma.$$

In the explosive event of pion decay, two photons are created, and they, in turn, may create new particles when they vanish.

The most convincing direct evidence of waves comes through the phenomena of diffraction and interference. *Diffraction* is the slight bending and distortion of a wave that occurs as it passes an obstacle. Long-wavelength radio waves, for example, bend around a building and can be detected in the building's "shadow." Particles would not be expected to be deflected just by passing near an obstacle. (Very short-wavelength radio waves behave more like particles and can be blocked by buildings.) If two beams of waves come together, they may show *interference*—"constructive" if the crest of one wave happens to match the crest of the other, and "destructive" if the crest of one wave matches the trough of another. (The physicist's definition of interference is broader than the ordinary definition. For waves, interference can help or hinder.)

It was through diffraction and interference effects that Thomas Young in England and Augustin Fresnel in France first clearly demonstrated the wave nature of light in the early part of the nineteenth century. Although we now regard their experiments as definitive, they did not change many minds at the time. Isaac Newton's particle theory of light still held sway.* But eventually the wave theory of light gained prominence—until the photon entered science and gave light a particle property after all. Then came the 1927 experiments of Davisson,

---

* The historians of science Gerald Holton and Stephen Brush have remarked that the particle theory of light was not as dogmatically entertained by Isaac Newton, its supposed champion, as by his later disciples. Even in physics, it is sometimes hard to shake hero-worship.

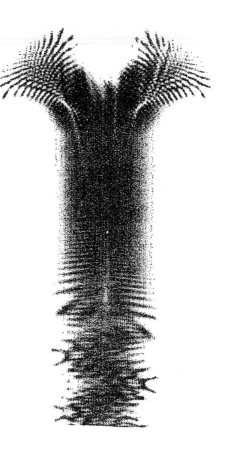

Figure 34. The shadow of a
screw shows diffraction and
interference.

Germer, and Thomson that revealed diffraction and interference effects
for electrons striking crystals. Since there is no reason to suppose that
particles (viewed classically) should diffract or interfere, these experiments, like those of Young and Fresnel more than a hundred years earlier, gave definitive evidence for wave behavior.

The modern version of the experiment that Thomas Young carried
out for light is called the double-slit experiment. The idea is relatively
simple (see Figure 35). A wave spreading out from a source impinges on
a sheet that is opaque except for a pair of narrow, closely spaced slits.
The portion of the wave passing through each of the slits diffracts,
spreading out on the far side of the opaque sheet. So at any point on a
detecting screen beyond the opaque sheet, waves will arrive from both

slits and can interfere, either constructively or destructively. In fact, they will do both, depending on the distances from the detecting point to the two slits. At the point exactly opposite the midpoint between the slits, the distance to the two slits is equal, so constructive interference will occur—crest to crest or trough to trough. Moving up from that point, a location will be reached where the distance to the upper slit is exactly one-half wavelength less than the distance to the lower slit. At that location, the trough of one wave will meet the crest of the other and destructive interference will result. Moving up still further, a location will be reached where the lower slit is one full wavelength farther away than the upper slit. This will put the two waves back in phase, crest to crest and trough to trough, so constructive interference will occur. The overall result is that a series of light and dark fringes will appear on the detecting screen. They are the smoking-gun evidence for wave behavior. Moreover, from measurements of the spacing of the two slits and the separation of the fringes, it is possible to calculate the wavelength.

In the quantum era, there have been two significant extensions of the double-slit experiment. First, it has been performed with electrons instead of light; the observed fringe pattern has provided a wavelength measurement of electrons that validates the de Broglie equation (as the crystal-scattering experiments also did). Second, it has been performed with a source of light so weak that only one photon at a time is in the apparatus, and with equipment that includes—instead of a detecting

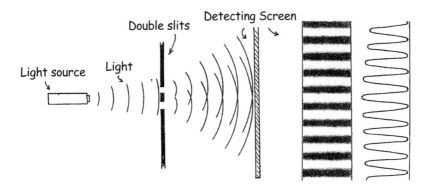

Figure 35. The double-slit experiment.

screen—an array of tiny detectors that can signal the arrival of single photons. This one-particle-at-a-time double-slit experiment, simulated in Figure 36, is one of the simplest yet most revealing demonstrations of all that is strange about quantum physics. Bohr and Einstein debated it in the 1930s, when it was still only theoretical. Later, with improvements in electronics and detectors, it was achieved in practice, and it continues to astonish scientists as well as nonscientists.

Here's what happens. Photon 1 is fired at the pair of slits. It is detected at a single point somewhere on the detector array. The point where it lands can't be predicted. Then photon 2, as nearly identical as possible to photon 1, is fired at the pair of slits. It, too, lands at some unpredictable point on the detector array. And so it continues, one photon after another. After ten photons have hit spots on the detector array, the pattern appears to be random. After a hundred photons have arrived, a pattern begins to emerge. Most of the photons are arriving at points where wave theory predicts constructive interference (most of these land in the "shadow" of the opaque screen). Fewer photons are arriving where wave theory predicts partial destructive interference. No photons reach points where wave theory predicts complete destructive interference. After a thousand or ten thousand photons have passed through the apparatus, a clear pattern of fringes is seen at the detecting array, exactly as expected for waves.

What is one to make of this? From what seem to be random and unpredictable individual events, a clear and simple pattern eventually emerges. You may be inclined to ask: Does a photon pass through one slit or the other? How does a photon "know" where it's OK to land and not OK to land? What determines its actual landing spot? Why is it that two photons, although in principle identical, never behave identically? The only explanation consistent with the observations, consistent with quantum theory, and consistent with an array of other experiments is that *each* photon individually acts as a wave as it makes its way from its source to a detector. Each photon passes through *both* slits. The photon is created at a point. It is detected at a point. Between its creation and its detection (really its annihilation), it behaves as a wave. *This is the essence of the wave-particle duality.* Moreover, probability enters the picture. No photon knows where it is to land. It knows only the proba-

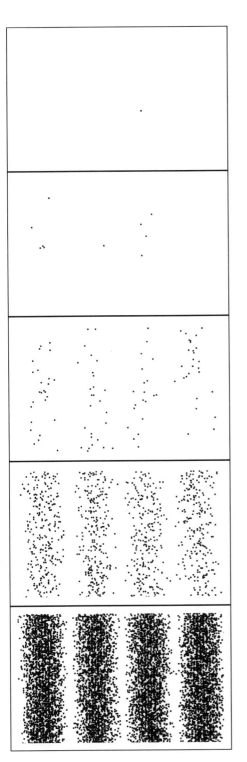

Figure 36. Simulation of the points where photons (or other particles) are detected after being fired one at a time at a double slit. The five panels show, respectively, the results for 1, 10, 100, 1,000, and 10,000 particles. Each panel includes the results of the previous panel. Images courtesy of Ian Ford; online at www.ianford.com/dslit/.

bility of landing at different points: high probability where wave theory predicts constructive interference, low probability where wave theory predicts destructive interference. (Think of tossing a coin a thousand times. Whether the coin lands heads or tails on a particular toss is unpredictable. Whether it lands heads or tails on a particular toss has nothing to do with how it has landed previously. Yet because of the working of probability, you can be reasonably confident that after a thousand tosses the coin will have landed heads about five hundred—but not *exactly* five hundred—times and tails about five hundred times.) A single photon, we have to say, can be diffracted—by two slits at once—and can interfere with itself. Whenever you look at a photon (that is, look at a detector, or actually register the photon on the retina of your eye), you see it at a point. When you aren't looking, it is a ghostly wave propagating through space just as electromagnetic waves do in classical theory.

Some physicists find the two-slit experiment so unsettling (the more you think about it, the more your head swims, to paraphrase Niels Bohr) that they consider quantum mechanics, despite its myriad successes and its absence of failures, incomplete. They believe that sometime in the twenty-first century—whether it be next year or fifty years from now—a new theory will come along, one that encompasses quantum theory and accounts for its successes but that makes more "sense." Even physicists untroubled by the oddities of the two-slit experiment are inclined to think that a reason for quantum mechanics has yet to be found.

## The Size of Atoms

For about a dozen years after J. J. Thomson's discovery of the electron in 1898, many physicists entertained the hope that a theory of atomic structure could be developed using only familiar classical concepts. One model of the atom advanced by Thomson himself was called the *plum-pudding model*. Thomson envisioned a spherical glob of positive charge (the "pudding"), in which were embedded tiny negative corpuscles (the electron "plums"). If the positive charge of the pudding matched the negative charge of the plums, the whole thing would be electrically neutral, and the atom's size would be dictated by the size of the glob of pudding. Theorists tried to figure out how the electrons might move around within the distributed positive charge, perhaps oscillating to and fro (which would help to explain why the atom emits radiation of only certain frequencies, like a musical instrument), and they tried to understand why such an atom might be stable. These efforts didn't really lead anywhere. Not surprising, you might think, given that the model is quite wrong. True—but often blind alleys must be pursued in science before the right road is found.

It took a crisis to illuminate the right road. The crisis was Ernest Rutherford's 1911 discovery that the atom is mostly empty space with a tiny nucleus at its center. Out the window went the plum-pudding model and other models of the time. Rutherford's new model was a beautifully simple planetary model, a small nuclear "sun" surrounded by electron "planets." Why was this a crisis? Because classical theory had something quite specific to say about how such an atom would behave, and it wasn't good news. Unlike the real planets of the solar system, which merrily orbit the Sun for billions of years, the electrons in the Rutherford atom would, according to classical physics, spiral into the central nucleus in roughly a hundred-millionth of a second ($10^{-8}$ s). The huge difference in behavior comes about because the electrons are electrically charged and are accelerating at a great rate. So, according to electromagnetic theory, they should emit radiation—of ever-increasing frequency—giving up their energy to that radiation as they spiral inward. This, of course, is not what atoms do. They retain their size and emit radiation only after being disturbed ("excited").

It was clear to Niels Bohr, when he learned of Rutherford's results, that quantum theory, which had recently been introduced to account for radiation, was going to have to play a role in the theory of matter—that is, atomic theory. In his 1913 paper, Bohr introduced ideas with staying power—discrete quantum states, quantum jumps, and angular-momentum quantization—but not yet the idea of waves. So Bohr's theory was an interim step toward explaining the size and the stability of atoms. I remarked earlier that Louis de Broglie, in advancing the idea of matter waves, saw how they could account for Bohr's rules and for the size of atoms. His breakthrough idea was that the size of an atom is dictated by the wavelength of the electrons within the atom. There has been only one major refinement of de Broglie's first thinking on the subject: we now know (as I mentioned earlier) that the waves are spread out over the three-dimensional volume of the atom—they don't just follow one orbital track. So the plums have become the pudding.

A wave, to be called a wave at all, must have at least one crest and one trough. It must rise and fall—perhaps repeatedly, but at least once. It can't be defined at a point. Its physical extension must be at least as great as its wavelength. So it is the wave nature of the electron, and specifically the wavelength of the electron, that determines the size of an atom. How does the electron decide whether to snuggle up close to the nucleus with a small wavelength, or range far from the nucleus with a large wavelength? Oddly enough, the answer is related to the reason that a marble, set rolling within a curved bowl, finally settles to the lowest point of the bowl. The marble seeks the state of lowest energy. So does the electron.

To understand the electron's search for the lowest energy, you have to think about two contributors to energy within the atom. First is the electron's *kinetic energy*. This energy increases as the electron's momentum increases, and therefore, according to the de Broglie equation, it increases as the electron's wavelength gets smaller. In other words, the more the electron wave pulls itself into a small knot close to the nucleus, the greater its kinetic energy becomes. If the electron were to follow classical expectations and spiral into the nucleus, its kinetic energy would grow without limit, because its wavelength would shrink to nothing. So the electron's search for the lowest energy is not advanced by diminishing its wavelength and diminishing the size of the atom. On the other hand, there is a second contributing energy: the *potential energy* associated with the attractive force between the nucleus and the electron. This energy *does* get smaller as the atom gets smaller, but at a lesser rate than the rate at which the kinetic energy grows larger. The two kinds of energy are, in effect, competing with each other. Because of its wave nature, the electron wants to spread itself over as large a volume as possible, so that its wavelength will be large and its kinetic energy will be small. It's almost as if the electron is being repelled by the nucleus.

Yet at the same time it is being attracted to the nucleus by electric force, and wants to shrink into a small volume. There is a certain atomic size at which the competing effects strike a balance and the total energy is minimized. This size proves to be about $10^{-10}$ m, quite large by the standards of the particle world.

As you might expect, Planck's constant, $h$, plays a role in fixing the size of atoms. If $h$ were smaller (that is, if quantum effects were less pronounced), atoms would be smaller. If $h$ were larger (if quantum effects were more pronounced), atoms would be larger. If $h$ were, hypothetically, zero, there would be no quantum effects. Electrons would follow the rules of classical physics and spiral into the atomic nuclei. There would be no atomic structure and no scientists around to ponder the matter.

It turns out that all atoms, from hydrogen to uranium and beyond, have about the same size. This can be accounted for by the same sort of energy competition I described above. If there were only a single electron around the uranium nucleus, that atom would be some ninety-two times smaller than a hydrogen atom. The point of balance between the electron's kinetic energy and the much stronger potential energy of the highly charged nucleus would be found closer to the nucleus. Because of the stronger attractive force, the electron's wavelength can be smaller and its kinetic energy larger than in the hydrogen atom. Indeed, the innermost electrons in the uranium nucleus do occupy a small region, covering only about $10^{-12}$ m instead of $10^{-10}$ m. But as more and more electrons are added to this atom, they spread over larger domains. The ninety-second and final electron joins a party in which there are ninety-two positive charges and ninety-one negative charges. It experiences a net charge of just one unit, the same as in hydrogen, and so settles into a state of motion similar to that of the lone electron in the hydrogen atom.

## Waves and Probability

It isn't hard to see how waves and probability might be linked. A wave is spread out over some region of space. Probability, too, can be spread out. An electron might be here or it might be there. Within an atom, for instance, an electron can't be said to be at any particular location (because of its wave nature), yet there are ways to probe and—with a certain probability—to find an electron at a particular point. In the hypothetical experiment shown in Figure 37, a high-energy gamma-ray photon is fired at an atom. It interacts with an electron, causing the electron (more exactly, *an* electron) to shoot out, along with a lower-energy photon. The two emerging particles, electron and photon, have large enough momenta that their wavelengths are much smaller than the size of the atom. This means that their paths are rather well defined and can (in principle) be traced back to a small region within the atom where the interaction occurred. The experimenter can say, "There must have been an electron at point P because that's where the interaction occurred." (Pinpointing a tiny region within an atom in this way is not actually practical. That's why this experiment is hypothetical.)

When this experiment is repeated, the point of interaction will be somewhere else within the atom. When it is repeated a thousand or a million times, a pattern will emerge. There will be certain regions in the atom where there is a high probability for the interaction to occur, other regions where the probability of interaction is low, and certain places— far from the nucleus, say—where the probability of interaction is negligible. The experimenter, before conducting the experiment yet again, will be able to say, "Although I have no idea where the interaction will actually occur on the next try, I know the probability that it will occur

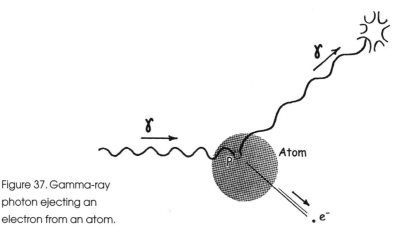

Figure 37. Gamma-ray
photon ejecting an
electron from an atom.

in any given region." So an electron that can be described as a spread-out wave within the atom can nevertheless interact at a point, and the probability of such interaction is linked in a simple way to the amplitude of the electron wave.

To give one other example, the wave of an alpha particle within a heavy radioactive nucleus has a tiny "tail" that extends well outside the nucleus. Because of this remote bit of wave, the alpha particle has a certain probability of springing through the force barrier that would otherwise hold it within the nucleus, whereupon it flies away to signal an alpha-decay event. This is the tunneling phenomenon, inexplicable in classical physics but explained in the quantum world through the wave nature of matter and the link between waves and probability.

Let me now turn back the clock. In March 1926, not many months after the young Werner Heisenberg had introduced his version of quantum mechanics, Erwin Schrödinger published his paper on wave mechanics. Schrödinger, an Austrian who was then a physics professor in Zurich, used his new wave equation to account for the known energy states of the hydrogen atom.* Like Heisenberg's work the previous year, Schrödinger's paper captured the immediate attention of physicists, who recognized its importance even though they didn't yet understand the meaning of Schrödinger's wave function.

It may seem strange that Schrödinger could successfully use his equation and that other physicists would applaud his achievement before anyone knew the meaning of the principal mathematical variable in his equation, the wave function. The reason is that the Schrödinger equation is a so-called *eigenvalue equation*.† It has "sensible" solutions

---

* Here is what the time-independent form of Schrödinger's equation looks like for a particle moving in one dimension:

$$d^2\psi/dx^2 + (2m/\hbar^2)[E - V]\psi = 0.$$

In this deceptively simple-looking equation, $x$ is the position coordinate along the direction of motion, $m$ is the mass of the particle, $\hbar$ is Planck's constant divided by $2\pi$, $V$ is potential energy, $E$ is total energy, and $\psi$ is the wave function. It is reliably reported that Schrödinger came up with this equation at the end of 1925 while on a winter holiday in Arosa, Switzerland, with a woman friend.

† The word "eigenvalue" is an amalgam of German and English. Such is the international character of physics.

only for certain values of the energy variable, $E$. Other solutions, for other values of $E$, are nonsensical and deemed nonphysical. So Schrödinger could find the permitted quantum energy states in hydrogen just by requiring that his wave function, whatever it might mean, behaved in a mathematically acceptable way (not becoming infinite, for instance).

Max Born, then forty-three and a senior physicist at Göttingen, had wide-ranging interests in theoretical physics. With his colleague Pascual Jordan, he had just made sense of Heisenberg's quantum mechanics by showing that it could best be displayed using the mathematics of matrices. Now he turned his attention to Schrödinger's new equation, and within three months came up with an idea that shook the already reeling world of physics. First, he said, Schrödinger's wave function, $\psi$, is an unobservable quantity. This was a startling new idea in physics. Until then, every concept that physicists dealt with in their equations described observable quantities. Second, he said, the *square* of the wave function, $|\psi|^2$, is the observable quantity and it is to be interpreted as a probability.*

To clarify this link between the wave function and probability, I will discuss the lowest-energy state of the hydrogen atom. The wave function of the electron in this state peaks at the location of the nucleus and falls "gradually" toward zero, becoming extremely small at distances beyond $10^{-10}$ m from the nucleus (see Figure 38). There are two ways to interpret this wave picture of the atom. First, we can say that the electron is not localized. It is nowhere in particular. It is spread over the entire interior of the atom, and its wave function shows how it is spread out. This statement focuses on the wave nature of the electron. Probability enters when we consider the particle nature of the electron. Although the electron is indeed spread out and is simultaneously everywhere within the atom, it always has the possibility of manifesting itself as a particle. If a high-energy gamma-ray photon arrives, as in the hypothetical experiment I discussed above, an interaction can occur at a point and an electron can shoot out from that point. The probability

---

* To repeat a point made in a footnote in Chapter 7: the wave function $\psi$ can be a complex quantity—that is, it can be represented by a number with both real and imaginary parts. It is the absolute square of $\psi$ that is to be interpreted as probability.

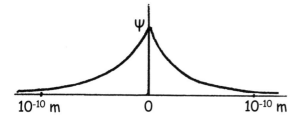

Figure 38. Electron wave function in lowest-energy state of a hydrogen atom.

that it will do so is proportional to the square of the wave function. Born himself said that he was led to his probability interpretation by realizing that although the electron does have a wave character, as de Broglie had said and as Schrödinger's equation implied, it surely has a corpuscular character too, showing itself often as a particle. So the wave-particle duality and the wave-probability link go hand in hand.

Even in early studies of radioactivity around 1900, there was evidence that nature's fundamental laws are probabilistic. And Rutherford's concern about Bohr's paper (how does the electron decide where to jump?) hinted that probability might play a role in quantum jumps. Yet physicists were not eager to replace certainty with probability, and preferred not to confront the issue head-on until Born forced them to. The implications of his conclusion go far beyond the way in which a wave function relates to a probability of *location*. Suddenly it became clear that all aspects of quantum behavior are probabilistic: the *time* when an event takes place, the *choice* among results of an event, and so on.

As I mentioned in Chapter 6, Einstein was unhappy about the appearance of probability in fundamental physics, and often said he could not believe that God played dice. Nearly all physicists have made peace with probability, but some remain uneasy about it. I very much doubt that the last word on quantum physics has been spoken.

## Waves and Granularity

Far from the center of an atom, the wave function of one of the atom's electrons is very close to zero. The wave function may rise to a single maximum at the atom's center, then fall again to near zero on the other side of the atom. Such is the wave function's behavior for the lowest-en-

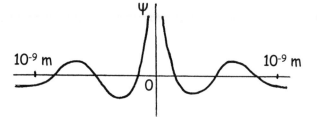

Figure 39. Electron wave function in higher-energy state of a hydrogen atom.

ergy state (the ground state) of the hydrogen atom, depicted in Figure 38. For more energetic states (excited states), the wave function undergoes two or more cycles of oscillation: either in and out, toward and away from the atom's center, as shown in Figure 39, or round about, circumferentially around the nucleus.

This behavior is reminiscent of the vibration of a violin string in its fundamental mode and in its higher harmonics. The lowest vibration frequency of a string held fixed at its two ends has one-half wavelength over the length of the string. The second harmonic is one in which there is a full wavelength over the length of the string. The third harmonic has one-and-a-half wavelengths from end to end. And so on. It is in the nature of the rules governing sound that the frequency of the second harmonic is twice the frequency of the fundamental (up one octave, in musical terminology), the frequency of the third harmonic is three times the frequency of the fundamental, and so on. When a violin string is bowed, all of these overtones, as they are called, sound together. The particular quality of the sound depends on the relative intensities of the overtones. But what is important for this discussion is that the frequencies of vibration of the violin string are quantized. A string of given length and tension vibrates only at a discrete set of frequencies.

As I mentioned earlier in this chapter, this classical quantization of wavelengths and frequencies is one of the things that set de Broglie thinking about the possibility of matter waves. Perhaps, he thought, matter waves—like waves on a violin string (or in the air column of a wind instrument)—could vibrate at only certain frequencies, and this fact would explain the energy quantization encountered in atoms. He was close to the mark. Physically acceptable solutions to Schrödinger's wave equation are those for which the wave function (and the probability) rapidly approach zero far from the atomic nucleus. Such solutions

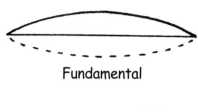

**Fundamental**

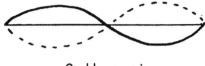

**2nd harmonic**

**3rd harmonic**

Figure 40. A vibrating violin string.

occur only for selected (quantized) energies. For other energies, the wave function becomes infinite where it should be zero. Nature doesn't avail itself of such a possibility. So the condition that the wave function be zero (or practically zero) at some distance away from the nucleus on both sides of the atom is exactly like "pinning down" a violin string at its two ends. The wave function, like the wave on the violin string, can have only certain wavelengths. The longest wavelength—a single rise and fall from one side of the atom to the other—corresponds to the fundamental vibration of a violin string and defines the ground state of an atom. Wave functions with more cycles of oscillation correspond to overtones and go with excited states of the atoms.

It may seem a little paradoxical at first that the wave function—though distributed continuously over space, in contrast to the discreteness of a particle—nevertheless accounts for the discreteness of energy. The musical analogy may help to make this reasonable.

Physicists like simple models, and in the quantum world no model is simpler than the *particle in a box*. Imagine an electron (or any other particle) constrained to move back and forth along a straight line between two perfectly impenetrable walls. In classical physics, it can be set mov-

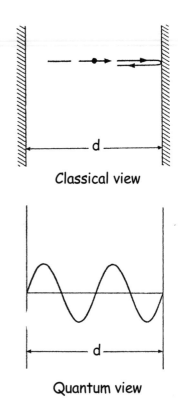

**Classical view**

**Quantum view**

Figure 41. Views of a particle in a box.

ing with any amount of kinetic energy, and it will continue to bounce back and forth with that energy indefinitely. The quantum description is very different. The particle has a wavelength determined by its momentum—short wavelength for large momentum, long wavelength for small momentum. For some arbitrarily chosen wavelength, the wave, snaking back and forth between the walls, will interfere with itself and soon wipe itself out. After a great many trips back and forth, the probability of finding the particle anywhere is zero. To avoid this calamity, the wavelength must be chosen so that the wave reinforces itself in its multiple trips between the walls. Figure 41 shows an example in which the space between the walls encompasses exactly two wavelengths. A wave of this length remains robust and self-reinforcing after any number of round trips. This choice defines an allowed, quantized energy for the particle in the box.

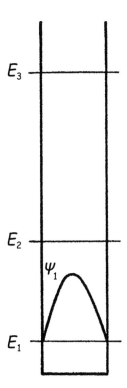

Figure 42. Energy ladder for a particle in a box.

The longest possible wavelength for the particle in a box is twice the spacing between the walls; that is, there's half a wavelength between the walls, just as there's half a wavelength in the fundamental tone of a violin string. The second possibility is one wavelength between the walls. Then 1.5 wavelengths are fit into the space, then 2.0, then 2.5, etc. These wavelengths get successively shorter, in a regular pattern. The momenta, being proportional to the inverse of the wavelengths, grow in a regular sequence—in fact, in equal steps. That is, the second state has twice the momentum of the first state, the third state has three times the momentum of the first state, and so on. If the particle is nonrelativistic (moving slowly relative to the speed of light), its energy is proportional to the square of the momentum. So the quantized energies of the particle in a box march upward with spacings that get larger and larger. Figure 42 shows the first three such energy levels.

The lesson of the particle in a box is that confinement leads to en-

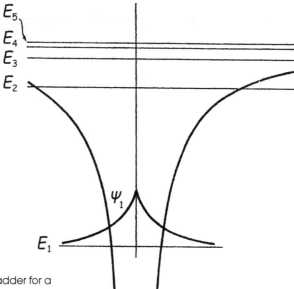

Figure 43. Energy ladder for a hydrogen atom.

ergy quantization. The closer together the walls are, the farther apart the energy levels. (Again, there is a musical analogy. The strings of a vi-olin vibrate at a higher frequency than those of a cello, which in turn vibrate at a higher frequency that the strings of a bass viol.) If a particle is completely unconfined, it may have any energy whatsoever, because it can have any wavelength whatsoever.

Now let's return from the idealized situation—the particle in a box —to the reality of an electron in an atom. The atom provides walls of a sort—walls of electric force. The atomic walls differ in two ways from the idealized walls. First, they are not impenetrable. The electron's wave function doesn't reach precisely zero at some point and remain zero out-side that point. It falls gradually (though in truth quite rapidly) to zero beyond the wall. Second, the atomic walls get farther apart as the elec-tron's energy increases. So an electron in an excited state is less con-fined than an electron in its ground state. The result of this is that the quantum energies in the atom actually get closer together as the energy increases, rather than farther apart, as they do in the case of the particle in a box. The difference is shown in Figure 43.

## Waves and Nonlocalizability

As I have said, a wave is, by its very nature, spread out. It extends over some region, large or small, depending on its wavelength and on what is confining it. The size of atoms can be understood as a consequence of the nonlocalizability of waves. The wave of a particle with a certain energy needs a certain minimum amount of elbow room, and this fact dictates the size of atoms. Another consequence of the nonlocalizability of waves is a limitation on how well the physicist can probe small dimensions. To get at a very small dimension means using a very small wavelength, and that, in turn, means a large momentum and a large energy. So physicists build gigantic accelerators, at gigantic expense, to study the smallest subatomic processes—an irksome reality of the nonlocalizability of waves.

Suppose you wanted to study ships in a harbor by analyzing waves that pass them. Waves rolling past a large ship at anchor would be strongly affected (as shown on page 12). The ship would leave a "shadow" of calm water, and the waves rounding the ends of the ship would be diffracted in a characteristic way. Some of these diffracted waves might interfere with one another. You could learn the size and shape of the ship fairly accurately by studying waves that had passed it in various directions. If, on the other hand, the same waves were to pass a piece of piling sticking out of the water, they would scarcely be affected and would show at most that some small thing was there, without revealing its size or shape. But you would have no trouble analyzing the piling with light waves—that is, by looking at it. The lesson: waves are an excellent tool of analysis if the wavelength is much smaller than the thing being analyzed. If the wavelength is much larger than the thing being analyzed, no details can be revealed. The wave has a certain "fuzziness" dictated by its wavelength.

The first experiments that revealed something about the inner structure of the proton were conducted by Robert Hofstadter at Stanford University in the mid-1950s. I knew Hofstadter when I was a graduate student at Princeton University and he was an assistant professor there, involved in inventing and perfecting better methods of detecting high-energy particles. It was no doubt his good fortune that Princeton didn't offer him tenure, for when he moved to Stanford he gained access to what he needed: electrons pushed to an energy of 600 MeV by a linear accelerator. These electrons had a wavelength of $2 \times 10^{-15}$ m, short enough (just barely) to study the interior of the proton. This wavelength is roughly the diameter of a proton, some 100,000 times smaller than the size of the hydrogen atom. Hofstadter was a superb experimenter; moreover, he wasn't content to leave the theorizing to others. He himself analyzed with great care the results of the experiments he carried out, which involved bombarding hydrogen with electrons and studying the pattern of electrons deflected by the protons at the centers of the atoms. He literally opened up the proton, showing it to be an entity of finite size having an inner structure, with both its electric charge and its magnetic properties being spread over its interior. For this work, Hofstadter was awarded the 1961 Nobel Prize in Physics.

Stanford scientists later built the two-mile-long Stanford Linear Accelerator, which now pushes electrons to an energy of 50 GeV. These electrons have a wavelength of $2.5 \times 10^{-17}$ m, much smaller than the dimensions of a single proton or neutron. At Fermilab's Tevatron near Chicago, the 1-TeV (1 trillion electron-volt) protons have a wavelength of about $10^{-18}$ m. A still-larger accelerator, producing particles of still-smaller wavelength, is under construction.* Interestingly, at energies in the multi-GeV and TeV range, protons and electrons of the same energy have nearly the same wavelength. This is because their kinetic energy is so much greater than their mass energy that it doesn't really matter whether they have mass or not, or how much mass they have.

I must add that the drive for higher and higher energy is not aimed solely at obtaining smaller and smaller wavelengths. The energy itself is important, for it is from the kinetic energy of the particles that the masses of new, heavier particles are constructed. Note, for instance, in Table B.4 that the masses of the W and Z bosons are some 80 to 90 times the mass of the proton, and in Table B.2 that the mass of the top quark is about 180 times the mass of a proton. If more massive particles

---

* The "large hadron collider" (LHC) at CERN in Geneva will push protons to an energy of 7 TeV. These protons will collide with equally energetic antiprotons, making 14 TeV available for the creation of new mass.

are waiting to be found, even more energy will be needed to create them.

As it happens, large wavelength is not always a drawback in subatomic studies. Neutrons slowed to thermal energy (less than 1 eV) have wavelengths around $10^{-10}$ m, comparable to the size of an atom. The "size," or "fuzziness," of such a neutron is tens of thousands of times larger than the size of a nucleus. It cannot reveal any details of the inside of a nucleus, but it can "reach out" to interact with nuclei that, according to a classical calculation, it is missing by a wide margin. In doing so, it reveals specific features of the neutron-nucleus interaction at zero angular momentum. Moreover, if the neutron is slowed sufficiently, it can interact with more than one nucleus at a time and can be diffracted by the array of nuclei in a sample of material. This reaching out is also important in nuclear fission. A slow neutron can find a nucleus of uranium 235 and stimulate a fission event even if its trajectory would seem to bypass all the nuclei.

## Superposition and the Uncertainty Principle

In 1927, amid the great intellectual ferment swirling around the creation of quantum mechanics, Heisenberg offered his *uncertainty principle*: for some pairs of physical quantities, the measurement of one of the quantities to a certain precision puts a limit on how precisely the other quantity can be measured. To put it simply, quantum mechanics imposes a limit on the ability to know. Like the de Broglie equation and like two-slit interference of matter waves, the uncertainty principle captures an important part of the essence of the quantum world.

One form of the uncertainty principle can be written

$$\Delta x\, \Delta p = \hbar.$$

On the right side is the ubiquitous Planck's constant (here divided by $2\pi$), which turns up in every equation of quantum mechanics. Momentum is represented by $p$, and position (distance) by $x$. The $\Delta$ symbols are used here to mean "uncertainty of" (not "change of"): $\Delta x$ is the uncertainty of position; $\Delta p$ is the uncertainty of momentum. The product of these two uncertainties is equal to the constant $\hbar$.* Since $\hbar$ is, on the human scale, extremely small, $\Delta x$ and $\Delta p$ can both be, for all practical purposes, zero in the everyday world. There is no limit imposed by nature on how precisely the position and momentum of a large object can be measured. If, for instance, you wanted to specify a person's location to within the diameter of a single atom, that person's speed could (in principle) be measured to a precision of about $10^{-26}$ m/s. In the subatomic world, it's a different story. An electron that is known to be located somewhere within an atom (that is, having an uncertainty of position of about $10^{-10}$ m) has an inherent uncertainty of speed of a million meters per second ($10^6$ m/s).

The uncertainty principle has captured the public imagination and is (regrettably) often applied in fields other than science. Those who wish to attack science can cite it as proof that the "exact" sciences are, after all, not so exact. Indeed, it is a profound statement about nature and about our ability to know, but it can be viewed as just one more aspect of the wave nature of matter, in which case it seems considerably less mysterious.

Understanding the uncertainty principle requires coming to grips with a somewhat difficult but extremely important idea: a wave can be confined (partially localized) only if different wavelengths are mixed. This mixing is called *superposition*, and it's an essential feature of the quantum world.

Consider first a wave that is *not* mixed, a wave of one definite wavelength, as shown in Figure 44. If this is a matter wave, it represents a particle moving along at constant speed, its momentum given by the (slightly rewritten) de Broglie equation, $p = h/\lambda$. Where is the particle? It is everywhere—or it is equally likely to be anywhere (along the infinite dimension of the wave).† Its uncertainty of position is infinite;

---

* Strictly speaking, this equation puts a *lower limit* on the product of uncertainties, the limit imposed by nature itself. Larger uncertainties can result from imperfections of measurement.

† At an instant of time, the square of the sine wave in Figure 44 has hills and valleys, so that the particle has a greater probability of being in some places than in others. But over time, as the wave propagates, the regions of high and low probability move. Averaged over time, the particle has an equal probability of being anywhere.

Figure 44. A "pure" wave of one wavelength.

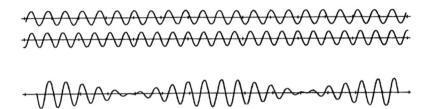

Figure 45. Two waves differing 10 percent in wavelength, separate and superimposed.

Figure 46. Superposition of many waves spanning 10 percent in wavelength.

its uncertainty of momentum (and of wavelength) is zero. This is an extreme situation that is at least consistent with the uncertainty principle. If the product of two uncertainties is constant, one of them can become vanishingly small only if the other becomes infinitely large.

Consider now what happens when several different wavelengths are superposed. Figure 45 shows the result of superposing just two waves that differ in wavelength by 10 percent. Five cycles from the point of maximum reinforcement (constructive interference), the waves interfere destructively. Five cycles further on, they interfere constructively again. The result is partial localization, the bunching together of the wave into regions each about ten wavelengths in extent.

If many different wavelengths are superposed but still spanning a total range of 10 percent, the result, as shown in Figure 46, is a bunching into just one region, some ten wavelengths in extent. In this case, the particle's uncertainty of position has been reduced to about ten wavelengths, and its uncertainty of momentum spans a range of 10 percent, the same (percentagewise) as the range of mixed wavelengths.

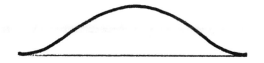

Figure 47. Result of superposing waves of many wavelengths.

Localizing the wave even more is possible. Figure 47 shows a wave that is collapsed down to a single rise and fall. Achieving this has required a nearly 100-percent range of superposed wavelengths and therefore a nearly 100-percent uncertainty of momentum.

As these examples show, squeezing down the uncertainty in a particle's position (small $\Delta x$) requires mixing many wavelengths, with a consequent large uncertainty in momentum (large $\Delta p$). The wave function of the ground state of the hydrogen atom illustrates the idea. The electron is confined within the atom, but one can't say exactly where. Its uncertainty of position is as large as the whole atom. Correspondingly, its momentum can be known only roughly. Its uncertainty of momentum, according to the uncertainty principle, can be calculated as $\Delta p = \hbar/\Delta x$.

Another form of the uncertainty principle that finds application in the particle world is

$$\Delta t\, \Delta E = \hbar.$$

Here, $t$ is time and $E$ is energy. Time and energy, like position and momentum, are a pair of quantities that cannot be simultaneously measured to arbitrary precision. If the uncertainty of time is small, the uncertainty of energy is large, and vice versa. As I described in Chapter 3, this form of the uncertainty principle can be harnessed for a useful purpose: measuring the lifetime of extremely short-lived particles. The short lifetime of such a particle means that its $\Delta t$ is small, so its $\Delta E$ is large. The experimenter measures the "spread" of energies observed in many decay events, thus determining $\Delta E$, from which $\Delta t$ and the particle's lifetime can be calculated.

On the other hand, the time-energy form of the uncertainty principle is not always helpful. The physicist, questing endlessly for ever more accurate timekeeping, wants to know to extraordinary precision the frequency of radiation emitted by certain quantum jumps in an atom. Since the frequency is determined by the energy of the transition through the Planck-Einstein formula $E = hf$, whatever uncertainty there may be in the energy of the transition will be reflected in an uncertainty of emitted frequency. The energy uncertainty, in turn, depends on the lifetime of the excited state. A state that lives a long time before it decays with the emission of a photon has a large uncertainty of time; the uncertainty in the energy of the emitted photon will be, correspondingly, small. The quest for the ultimate in timekeeping becomes, in part, a quest for suitable excited states in atoms with very long lifetimes.

To further peel back the mystery of the uncertainty principle, I should note that a time-frequency version of the principle is known in the large-scale world. It doesn't involve Planck's constant but it does involve waves. An electrical engineer knows that transmitting a tone of precise frequency over a wire is impossible unless the tone extends over a relatively long time. Shortening the pulse introduces, willy-nilly, an uncertainty of frequency—that is, a range of frequencies. The audible tones transmitted when you make a telephone call, for example, can't be sent at a lightning pace. If the time of each tone were too much reduced, the frequencies would get mixed up and the receiving station couldn't tell which button you had pushed. (An automatic dialer may speed up the "button pushes" to ten per second or thereabouts, but seeking greater efficiency with a thousand pulses per second would be self-defeating.) And an organist knows that it's unwise to try to play very low-frequency notes at too brisk a pace. The tones would sound muddy due to frequency mixing if the duration of each tone were excessively reduced.

The all-important difference between Heisenberg's uncertainty principle and the uncertainty exhibited by waves in the everyday world is the de Broglie link between wavelength and momentum. That is a purely quantum connection that brings in Planck's constant and has no counterpart in the classical world.

## Are Waves Necessary?

What's that flying through the sky? It's a wave! It's a particle! It's both! That's a typical way to speak of the wave-particle duality. Saying that something is both a wave *and* a particle is, to be sure, a mind-bending idea, hard to accept and absorb. But if you keep in mind that the particle aspect is evident under certain conditions and the wave aspect under other conditions, the wave-particle duality becomes less dizzying. Think of a man who is Casper Milquetoast at the office and Mad Max behind the wheel of his car. You wouldn't say that he is both Casper and Max at the same time; rather, depending on circumstances, he can be one or the other. So it is with a quantum entity. It is a particle when it is created and annihilated (emitted and absorbed). It is a wave in the interval between those events. Still, adopting this point of view is not sufficient for overcoming the impression of quantum weirdness. You may still want to ask: If the wave starts to propagate outward from the point where the particle is created, how does the wave know where and when to "collapse" to signal the annihilation (the detection) of the particle? The only answer to this question is that the wave is a wave of probability, of potentiality. It gives the likelihood that the particle will end its life at some future time and some other place.

In 1940 or 1941, as a graduate student at Princeton University, the always effervescent Richard Feynman went to his research advisor, John Wheeler, and said, in effect: Who needs waves? It's all particles. Feynman had a new vision of a quantum entity's history between the moments of its birth and death. He discovered that he could get correct quantum results by replacing the wave by an infinite number of particle paths. From the point of its creation, the particle follows—simultaneously—all possible paths to any given distant point. Each path has a certain "amplitude," and these amplitudes all add up to make possible a prediction of the probability that the particle will in fact be detected at that point. (It could happen, for instance, that the amplitudes, some positive and some negative, add to zero, yielding zero probability that the particle will land at a particular point, such as in a dark band in an interference pattern.) Wheeler, entranced by the idea, gave it a name, *sum over histories*, and immediately made an appointment to go and tell Einstein about it. Recalling that meeting later, Wheeler wrote:*

---

* John Wheeler, *Geons, Black Holes, and Quantum Foam: A Life in Physics* (New York: Norton, 1998), p. 168.

I was excited by the idea. I went to Einstein at his home and sat down with him in his upstairs back room study, spending about twenty minutes telling him about Feynman's idea. "Professor Einstein," I concluded, "doesn't this new way of looking at quantum mechanics make you feel that it is completely reasonable to accept the theory?"

He wasn't swallowing it. "I still can't believe that the good Lord plays dice," he answered. . . .

Einstein, as he once said about himself, could be as stubborn as a mule.

Feynman's provocative way of thinking didn't really banish waves. He merely provided an alternative way of looking at quantum phenomena. But it was a very important alternative because he emphasized the superposition of multiple amplitudes, a central tenet of quantum mechanics. Feynman also illuminated what might "really" be happening as those evanescent waves spread around. In many applications, the wave point of view remains simplest to use. Even with the sum-over-histories approach, concepts such as wavelength, diffraction, and interference show up.* So the answer to the question posed at the beginning of this section, "Are waves necessary?" is: "No, not really." But what is going on mimics wave behavior so closely that one might as well use waves to describe reality.

---

* In his charming book *QED* (Princeton, N.J.: Princeton University Press, 1985), Feynman sets forth the sum-over-histories approach to the behavior of photons and electrons.

# chapter 10

# **Pushing the Limits**

Three questions can be asked about the future of quantum physics. First, will scientists and engineers harness the quantum world? That is, can the laws and the phenomena of the subatomic world be put to practical, technological use in society? One answer is that they already have been—through lasers, microcircuits, scanning tunneling microscopes, and nuclear reactors (not to mention nuclear bombs). Another answer is that new and startling applications, such as quantum computing, are likely. And a hypothetical, if far-fetched example worth discussing is the use of particle-antiparticle annihilation as motive power for space transportation (or as explosive power for ultimate bombs).

The second question: Will new riches of understanding emerge in what might be called the sub-subatomic world—that is, at distances vastly smaller than any studied so far? There are hints that this may be the case, and that in the super-small domain quantum theory and gravitation will be united.

And the third question: Will a reason for quantum theory be discovered? For seventy-five years, the theory has made physicists uneasy, even though it has never failed an experimental test. For years, Niels Bohr got up every morning to do battle with quantum theory, and he and Albert Einstein never tired of debating it. After a heart attack in 2001, at age 89, John Wheeler said, "I may not have much time left. I'd better spend it thinking about the quantum." And a growing crowd of younger physicists are busy doing just that.

Why are physicists troubled by a theory that works flawlessly? They are uneasy not just because it violates intuition and common sense. Relativity does that, too, and it troubles no one. They are uneasy because quantum theory deals with ghostly, unobservable quantities (wave functions), makes probability fundamental, and leaves the boundary between the quantum realm and the realm of human perception quite vague. The reason for quantum theory, if there is one, could come from below or from above. That is, it might be found in the sub-subatomic domain of the smallest intervals of space and time, or it might be found in some cosmic principles governing the universe at large.

I hope that answers to all three questions will be found in the lifetime of my younger readers. If found, the answers will push the limits of technology and intellect.

## Quantum Physics and the World We Inhabit

When you first think about it, the quantum world and the classical world seem quite remote from each other. In everyday life, you don't see single photons; you never watch a single atom jumping from one quantum state to another or a pion suddenly disappearing where a muon and a neutrino pop into existence; you are not aware that there are any limits, in principle, to how accurately you can measure something; and you have never seen a baseball pass through two slits at once or exhibit an interference pattern because of its wave nature. Indeed, the remoteness of quantum phenomena from everyday experience explains why the quantum theory is a scientific newcomer in human history and is part of the reason that quantum phenomena seem so strange. On some planet in a remote galaxy where little creatures are directly aware of quantum phenomena, quantum theory would have been discovered early in the history of that planet's science. But would these creatures find quantum phenomena to be perfectly normal and sensible or would they, like Bohr, forever ponder its meaning?

When you think further, you realize that just about everything in our everyday world is the way it is because of quantum mechanics. Matter has bulk because atoms have size, and atoms have size because of their quantum nature. The colors, textures, hardness, and transparency of materials; the nature of substances at ordinary temperature (whether solid, liquid, or gas); the readiness or reluctance of an element to react chemically with other elements—all of these things depend ultimately on the exclusion principle regulating the behavior of electrons within atoms, and thus on the fact that electrons are fermions of spin one-half. The red light from a storefront's neon sign and the yellow light from a sodium vapor streetlamp are the colors they are because of specific quantum jumps within particular atoms. The interior of the Earth is hot because heavy elements undergo radioactive decay and release energy over billions of years, guided by the weak interactions, the penetration of "impenetrable" barriers, and the laws of quantum probability. And the sun shines because of a combination of strong, weak, and electromagnetic interactions that releases energy as the nuclei of hydrogen atoms gradually fuse into helium nuclei. The list could go on. Only when we get into the gravitational realm of satellite and planetary motion do the laws of quantum mechanics recede into the background.

Sometimes more "purely" quantum phenomena show themselves in the large-scale world. The Bose-Einstein condensate is an example. So are the superconductivity exhibited by certain materials at low temperatures and the superfluidity of liquid helium, also seen at low temperature.

You have no doubt heard that perpetual motion is impossible. It is inconsistent with the laws of thermodynamics. Patent examiners routinely deny patents for perpetual-motion machines without having to study the claims in depth, on the grounds that these inventions violate the laws of physics. But these are the laws of complex systems with interacting parts. Perpetual motion in the quantum world is (fortunately) commonplace.* An electron in an atom never gets tired. Friction never

---

* For all practical purposes, perpetual motion of astronomical bodies is also commonplace, although tiny frictional forces do eventually change the orbits and spins of the moon and planets.

slows it down. It just keeps going. Under very special circumstances, as in a superconductor or a superfluid, this frictionless perpetual motion that is common at the atomic scale manifests itself also at the human-sized scale. If quantum computers are to be realized, they, too, will have to take advantage of the frictionless character of subatomic motion.

## Using Antimatter?

Antimatter as a power source is possible in principle. For that reason, and because it's fun to talk about, I will devote some space to it here. It is surely "pushing a limit." If it ever becomes practicable, this will occur in some advanced civilization unlike our own. For us humans, it is a pipe dream.

Nothing—not antimatter or anything else—is truly a *source* of energy, for energy can only be transformed, not created or destroyed. When you "consume" energy, you are really transforming it from a more useful to a less useful form (and usually paying for the privilege). Yet it is commonplace (and handy) to talk about energy sources. In everyday usage, an energy source is either stored energy (as in gasoline or a battery—or, hypothetically, in antimatter) or energy in transit (as in solar energy or wind). Among the concepts of physics, energy is the most multifaceted, so there is a rich variety of energy transformations that find practical use.

Energy that is put to use may have been stored for only a little while or no time at all, as in the wind that drives a sailboat. Or it may have been stored for dozens of years, as in the wood that burns in a fireplace. Or for millions of years, as in the coal that fuels a power plant. Or for billions of years, as in the uranium driving a nuclear reactor (its origin being supernova explosions long ago). At the outer limit of storage time is the hydrogen that powers the Sun, dating from soon after the Big Bang some fourteen billion years ago.

Hydrogen used for its chemical (as opposed to its nuclear) energy is sometimes described as a nearly inexhaustible energy source because it is a main constituent of sea water. If only! In fact, because of inevitable inefficiencies, the energy needed to extract hydrogen from water (or from hydrocarbons) is more than the energy released when the hydrogen is burned or used in a fuel cell. Hydrogen is one more way to store and transport energy, useful because it is easy to pipe from one place to another and because whatever pollution it causes is located at the site where it is created (where it can be controlled), not at the site where it is used.

Antimatter in the world of science fiction is treated like hydrogen in the real world. It is made from commonly available material by expending energy; then it is stored, transported, and converted into useful energy where energy is needed. There is, indeed, no more potent "source" of energy than antimatter, so it is the natural choice to power the Starship Enterprise. When gasoline is burned in an automobile engine, less than one billionth of the mass of the gasoline is transformed into energy. When uranium nuclei undergo fission to power a reactor, about a thousandth of the mass of the uranium is transformed into energy. When antimatter and matter annihilate to power the Enterprise, 100 *percent* of the mass of the antimatter is transformed into energy (or 200 percent if you include the contribution of the annihilating matter). An anti-pea from an anti-peapod would release as much energy as half a million gallons of gasoline—enough to run a fleet of one thousand cars for a year. An anti-pea arranged to annihilate in a single burst of energy would be a bomb as devastating as the one that leveled Hiroshima in 1945.

How likely is all this? There is a tendency to think that anything that is possible in principle will sooner or later be realized in practice, if humans put their minds to the task of making it happen. In this case, I offer the opinion that it will never happen. Storing antimatter is an insurmountably difficult task. Antimatter was discovered in 1932 with the observation of the positron. Antiprotons were first created in 1955, and antineutrons in 1956. Since then, antiparticles have been routinely created and studied in high-energy accelerator laboratories, but always in infinitesimal quantities and always at such high energy that they have no inclination to form antiatoms. Researchers at Switzerland's CERN laboratory took a giant stride forward in 2002 by creating and storing millions of antiprotons and millions of antielectrons (positrons), from which they were able to form tens of thousands of antihydrogen atoms, and see the annihilation of about a hundred of them. As you can surmise, an antihydrogen atom consists of a negative antiproton surrounded by a positive antielectron. It is a fascinating object of study if you wish to check on matter-antimatter symmetries in nature, but is not a practical energy source. In the CERN experiment, the charged antiparticles were stored (temporarily) in magnetic fields so that they touched no matter. The neutral antiatoms, once formed, escaped from the magnetic trap and annihilated in the walls of the container.

Ten thousand antihydrogen atoms sounds like a lot. In terms of available energy, it doesn't amount to much. A dab of gasoline with the same potential energy as ten thousand antihydrogen atoms would be too small to see without a microscope. Even a billion antihydrogen atoms, upon annihilating, would provide only as much energy as you use in driving your car five thousandths of an inch forward. For a useful amount of energy, you would need more than a billion billion ($10^{18}$) antihydrogen atoms. The *Star Trek* crew never reveal how their fuel is stored.

Some things in nature that are possible in principle do not happen. The molecules of air in the room where you are sitting could, for instance, "decide" to cluster in some other part of the room, leaving you gasping for air. (If you are reading outside, the air molecules near your head could all flee at once, with the same unpleasant effect on your breathing). These things don't happen. The chance that they might happen can be calculated readily. For such a molecular clustering to occur even once in the lifetime of the universe is out of the question. The chance that we humans may figure out a way to store and later use a significant quantity of antimatter is not quite as remote as the chance that a statistical fluctuation will take your breath away, but it still seems to me to be beyond the possible.

Most antiparticles, like most particles, are unstable. That is, independent of their penchant for annihilation, they undergo spontaneous decay into other particles. This is true of antineutrons, antilambdas, antimuons, and so on. Among fermions, only antiprotons, antielectrons, and antineutrinos are stable. Since antineutrinos can't be corralled, that leaves only antiprotons and antielectrons as potential antimatter energy sources. If you want to be fanciful, you could say that antineutrons could also be stabilized by joining with antiprotons to form antinuclei. This would lead to an anti–periodic table.

Is there antimatter—are there perhaps even antigalaxies—elsewhere in the universe? This cannot be completely ruled out, but it seems unlikely. If there were antimatter regions, there would be boundary zones between the matter and antimatter parts of the universe, and in those boundary zones stray electrons and antielectrons would meet and annihilate, giving rise to pairs of gamma-ray photons of characteristic energy—about 0.5 MeV each. Such radiation has not been detected from intergalactic space. (Proton-antiproton annihilation would produce even more energetic radiation, but would be less easy to recognize because it comes in a range of energies via intermediate pions.)

But even if there is no "primordial" antimatter (left over from the Big Bang), there is a certain amount of it that is being created all the time by high-energy processes in the present universe. Some antiprotons are seen in the cosmic radiation that strikes the Earth, and radiation from annihilating positrons is seen coming from the center of our own galaxy.

According to presently accepted theory, for the first millionth of a second or so after the Big Bang,* the universe was a dense hot soup ("maelstrom" might be a better word) of quarks, leptons, and bosons (including photons). When the universe reached the age of a millionth of a second, give or take a bit, and the temperature had fallen to around ten trillion degrees, the quarks joined up in triplets to form not-quite-equal numbers of protons and antiprotons (also neutrons and antineutrons). The numbers of these baryons and antibaryons were not equal because of a tiny matter-antimatter asymmetry: the CP nonconservation discovered by James Cronin and Val Fitch in 1964 (see Chapter 8). As a result, for every billion antiprotons milling around, there were a billion and one protons, with a similar asymmetry for neutrons and antineutrons. By the time the expanding universe cooled to around one hundred billion degrees (within the next hundredth of a second or so), annihilation had eaten up all the antibaryons and, with them, almost all

---

* During this time an enormous and enormously rapid expansion of the universe occurred (the so-called *inflation*), but that's a story outside the scope of this book.

the baryons. (The erasing of the much less massive positrons came later, when the universe was some fifteen seconds old.)*

The present-day universe of galaxies, stars, planets, and you and me is made of that leftover one billionth. The fact that the universe now contains about a billion photons for every proton and seems to contain no appreciable number of antiprotons supports this theory. As Val Fitch puts it, the slight asymmetry of matter and antimatter is the reason we are here.

## Superposition and Entanglement

In Chapter 9 I discussed the superposition of waves of different wavelength that leads to localization (or partial localization) of waves. Now let me bring the de Broglie equation ($\lambda = h/p$) back onstage. It says that wavelength is tied to momentum: for every specific wavelength, there is a specific momentum. This means that superposing wavelengths is equivalent to superposing (or mixing) momenta. So what we call a single state of motion of an electron—say, in a hydrogen atom—can also be regarded as a mixture of many different states, each of a different momentum. *One* energy, *many* momenta.

What I have laid out in the previous paragraph could be called the signature property of quantum mechanics—the thing that, above all else, sets quantum theory apart from classical theory. In classical theory, for instance, an electron orbiting a nucleus has, at every instant, a definite energy and a definite momentum. The momentum changes from moment to moment, but the idea that the momentum might have two or more values at the same time is completely foreign to classical theory. The classical physicist would call such an idea simply nonsense. Yet superposition is the rule in quantum theory. *Every* state of motion of a particle or a nucleus or an atom can be regarded as a superposition (or mixing) of other states, sometimes even an infinite number of other states. That is much of the reason the quantum world seems so strange.

---

* If the total charge of the universe is zero (apparently true, but no one knows why), the number of surviving electrons conveniently matches the number of surviving protons.

If you ask, "What is the momentum of an electron at a particular moment as it moves in its lowest-energy state of motion in a hydrogen atom?" the quantum physicist answers, "It is a mixture of a vast number of different momenta." Suppose you persist and ask, "But can't you measure the electron's momentum and find out what it is?" Then the quantum physicist must answer, "Yes, I can—and if I do, I find a particular momentum. The very act of measurement selects one among the many mixed momenta." That is where superposition and probability join hands. If the measurement is repeated many times with many identical atoms, many different results will be achieved. The probability of any particular result is determined by the way in which the different momenta are mixed. They are stirred together with different "amplitudes," one for each momentum; and the square of each amplitude gives the probability that that particular momentum will be measured.* Now comes a very important point. Superposition does *not* mean that an electron may have one momentum or another momentum and we just don't know which it has. It means that the electron literally has *all* the momenta at once. If you can't visualize this, don't worry. Neither can the quantum physicist. He or she has learned to live with it.

Here is another example of superposition, one in which only two states are mixed. As shown in Figure 48, a certain electron has its spin directed to the east. We normally say that an electron's spin can point in only two directions, up or down. But any two opposite directions will do. So if this electron's spin is directed to the east, it is *not* directed to the west. Its spin direction is definite. We say that it is in a definite *state*, with east-directed spin. But this state, like every other quantum state, can be expressed as a superposition of other states. It can, for instance, be viewed as a superposition, with equal amplitudes, of north-directed and south-directed spin, as indicated in Figure 48. If an experimenter sets up a measuring apparatus to see if the spin is directed to the north, there is a 50-percent chance that the experiment will indeed find a north-directed spin. In other words, a spin that is known for *sure* to point eastward can, half the time, be measured to point northward (and

---

* As noted earlier, the probability is actually the *absolute* square of an amplitude that might be a complex number.

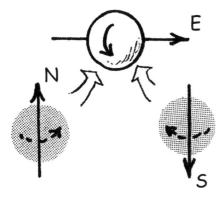

Figure 48. The east-directed spin is an equal mixture of north- and south-directed spins.

half the time to point southward). Numerically, the north-spin amplitude and south-spin amplitude are both 0.707, their squares being 0.50. At the same time, the east-spin amplitude is 1.00 and the west-spin amplitude is zero.

By now, you may be wondering in which direction your head is spinning. But it could be worse. If your head were an electron, it would be spinning simultaneously in two opposite directions, along any axis you wish to choose. The electron with its east-directed spin, for example, is also a superposition of spins directed to the northeast and to the southwest, as indicated in Figure 49. An experimenter who sets out to see if the spin is directed to the northeast will find that 85 percent of the time it is, but that 15 percent of the time it is not (assuming the experiment is repeated identically many times).

To summarize: any *single* state of a particle or a nucleus or an atom—or of any quantum system—is also, at the same time, a superposition (or mixing) of two or more other states. They are mixed in a strange way: not like the ingredients in a cake mix, which lose their individuality, and not like lanes of traffic, which retain their separateness. Quantum mixing (*superposition* is the technical name for it) inseparably joins the parts in what might best be visualized as intermingling, like overlapping waves from two pebbles dropped in a pond. The overlap is preserved so long as the system remains undisturbed and unmeasured. It's like a grand game of hide-and-seek. The moment the system is observed, or even

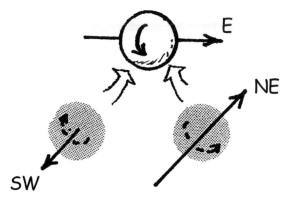

Figure 49. The east-directed spin is an unequal mixture
of northeast- and southwest-directed spins.

when it interacts with some "classical" object (not necessarily a human
observer), one component of the mixture is magically extracted while
the other components, equally magically, vanish. Which component
manifests itself is governed by probability—the *fundamental* probability
that is part of the quantum world.

Superposition gets really interesting when two parts of the system
become widely separated in space. In the next section I will discuss one
such example, the *delayed-choice experiment*, where a single photon exists as the superposition of two states that follow completely different
tracks. The photon, as it were, divides.

One may also have a situation in which two or more different systems become superposed. Such a case is the Bose-Einstein condensate,
where the different systems (atoms) are literally superposed in space.
But another possibility is that two "systems" (two particles, for instance)
are created together in a superposed state at one place and then fly
apart. For example, an experimenter could arrange for an electron and a
positron to annihilate into a pair of photons whose total spin is zero.
Since each photon has a spin of one unit, this means that the two photons, no matter how far they separate, must retain oppositely directed
spin. In one of the two states that are superposed, photon 1 has spin up
and photon 2 has spin down. In the other state, it is the opposite: pho-

ton 1 has spin down and photon 2 has spin up. The superposed state created when the electron and positron annihilate can be expressed as

$$(1 \text{ up}, 2 \text{ down}) + (1 \text{ down}, 2 \text{ up}).$$

The two possibilities are intermingled, or mixed, and remain so until one or the other photon has its spin measured or interacts with something else. An experimenter working five meters (or five light-years) from where the two photons were created measures the spin of photon 1 and finds it to be directed upward. At once this experimenter knows that the spin of photon 2 is directed downward. But the measurement could have yielded down spin for photon 1, in which case photon 2 must have up spin. This is the "spooky action at a distance" that so deeply troubled Einstein about quantum mechanics. Each photon, flying along its own separate path, has a superposition of possibilities, spin up and spin down. The act of measurement causes one of these possibilities to be realized and instantly reveals the direction of spin of the other photon—which, until that moment, was itself in a mixed state of spin up and spin down. Something very close to the experiment I describe here has actually been performed, with photons emitted by atoms rather than by particle annihilation, and with separation distances of meters, not light-years. But there is little doubt that if the photons could fly unimpeded for light-years, and if we could make a deal with aliens to help in the measurements, the results would also confirm the predictions based on quantum superposition.

When the superposition involves two or more systems that become separated in space, it is usually called *entanglement*. It's a good word. You can see that the states of the two photons flying apart are indeed entangled—a bit like family members whose lives are entangled no matter how far apart they dwell. But from a fundamental point of view, superposition and entanglement are really the same thing. The reason is that two superposed systems constitute a single system. There is no difference in principle between two superposed states of a single atom and two superposed atoms. The two photons flying apart in the example discussed in the previous paragraph are really parts of a single system. A single wave function describes their joint motion.

A system such as an electron whose spin can point in either of two directions has come to be called a *qubit* (pronounced "CUE-bit"). It is the analog of the binary bit that is the basic unit of computing. But there is an important difference. The classical bit is either on or off, either up or down, either zero or one. It is never both at once. The qubit *can* be both at once. Because of superposition, a qubit can exist as a mixture of on and off, or up and down, or zero and one. Moreover, it doesn't have to be an equal mixture. The qubit can be, say, 87 percent up and 13 percent down, or any other combination. Thus, it contains, in principle, much more information than a classical on-off, yes-no bit.

The properties of qubits have led in recent years to a great burst of activity in the new field of quantum computing (still entirely a theoretical field and a long way from practical realization). The guiding principle of quantum computing is that a suitably designed "logic gate" (similar in principle to what lies at the heart of every computer) can simultaneously process *both* the on and off, or zero and one, possibilities of a qubit, rather than processing just one or the other, as a classical logic gate has to do. This leads, theoretically, to a gain in processing power much greater than twofold. If one imagines a single system consisting of a number of superposed qubits, there can be a vast number of amplitudes describing all the ways that the qubits can be mixed. Two qubits can be mixed in four ways, ten can be mixed in a thousand ways, twenty can be mixed in a million ways. The logic gate, acting just once on this superposed system, can process all the possibilities at once—provided it can do so without disturbing the system, because any interaction would extract from all the possibilities just one, with a certain probability. So the quantum computing theorist must think about how a superposed (or entangled) system could survive a processor and emerge on the other side to be analyzed later.

Eventually, information must be extracted from the qubit or the entangled set of qubits. When this happens, bingo!—just one classical bit of information is extracted. The act of measurement has pulled out only one of the myriad possibilities that were, until then, superposed. Does this undo the theoretical advantage of the quantum computer? No, because in many problems, a single simple answer is all that is wanted. Although scientists working in this field don't think that weather forecasting is a likely future application of quantum computing, it does afford an example in which complex input is needed to achieve a simple output. You may ask, for example: Will it rain in Philadelphia tomorrow? The simple answer, yes or no, requires a vast amount of input data and a vast amount of calculating. In problems of this kind, quantum computing may come into its own. David Deutsch, a reclusive genius who seldom strays from his home in Cambridge, England, and who is a pioneer in quantum computing, has pointed out that the single bit of information finally extracted from a quantum calculation can be the result of interference among all the superposed amplitudes, so that the vast information in the superposition is, so to speak, funneled into the final answer.*

## Delayed Choice

John Wheeler, now an impish man in his nineties, likes to attribute his legendary courtesy and his unfailing friendliness to the fact that he was a near-sighted teenager. He says that as a Johns Hopkins student in the late 1920s, he wanted to greet friends whom he encountered while walking across the campus. But because of his poor vision, he couldn't tell whether he was approaching a friend or a stranger. To be safe, he gave a smile and sometimes a friendly wave to everyone he passed. It became a habit.

Whatever Wheeler's limitations of eyesight, his vision in a larger sense is strikingly unimpaired. He is as renowned for his ability to look beyond the horizon of physics as he is for his personal warmth. For example, he postulated the existence of *geons* (concentrations of photons so intense that they circle around a common center, held in orbit by their own gravity without any material substance—entities still unobserved); he named and explored black holes before most physicists believed in their reality; he introduced the so-called Planck scale of incredibly small distances and short times, where quantum uncertainty acts on spacetime itself to create "quantum foam" (another Wheeler coinage); and through his aphorism "It from bit" (the idea that the real

---

* For more on quantum computing and on David Deutsch, see Julian Brown, *The Quest for the Quantum Computer* (New York: Simon and Schuster, 2000).

John Wheeler (b. 1911), ca. 1980. Photo courtesy of John Wheeler.

world—"it"—might rest ultimately on information, or "bits"), he set the stage for all of the current work on quantum information theory. In this section I will describe an experiment that was beyond the horizon when Wheeler proposed it in 1978, but that has been realized in several laboratories since, beginning with the work of Carroll Alley and his collaborators at the University of Maryland in 1984.* It is an arrangement that allows an experimenter to decide, long after a photon has left its source, whether the photon should follow a single track to a detector or should proceed along two superposed tracks.

To describe the experiment, I am going to borrow the same baseball diamond that Wheeler used in his 1998 autobiography, *Geons, Black Holes, and Quantum Foam*. If you are not familiar with American baseball, please find someone who can help you to visualize what is going on. I will pretend that I am the experimenter. Behind home plate and a little off to one side, I have set up a light source that directs a stream of photons toward a half-silvered mirror mounted above home plate. This is a glass plate bearing a layer of silver so thin that it reflects only half the light that hits it, and lets the other half through. This means that an individual photon has a 50-percent chance of being reflected and a 50-percent chance of being transmitted. You can guess what else it means: that a photon hitting the half-silvered mirror will be *both* reflected *and* transmitted, its wave function dividing. After hitting the mirror, it will be in a superposed state of two different propagating directions.

I set up the half-silvered mirror so that the light passing through it heads toward first base and the light reflected from it heads toward third base (see Figure 50). Above these two bases I place fully reflecting mirrors that direct all light (or all photons) toward second base. If I leave the space above second base empty and place detectors in left field and right field, these detectors will register the arrival of photons (*whole* photons!). A click in left field means that a photon has arrived via first base. A click in right field means that a photon has arrived via third base. So far, there is no particular evidence of superposition—merely a measure of probability for the two separate paths. On average, half of

---

* A rich variety of experiments involving entanglement and delayed choice are under way now in laboratories in Paris, Geneva, and Vienna.

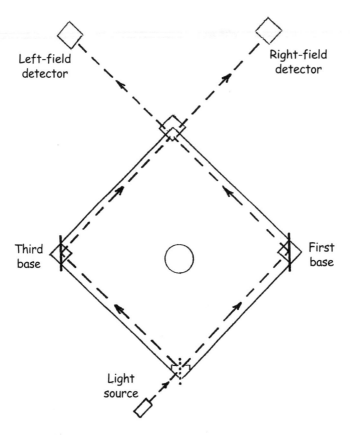

Figure 50. With nothing above second base, each photon has a 50-percent chance of arriving in left field and a 50-percent chance of arriving in right field. So half of the photons arrive at each detector.

the photons will be detected in left field and half will be detected in right field.

To demonstrate that the photons are really following two paths at once, I find another half-silvered mirror and place it above second base, so that half the light arriving from first base is reflected into right field and half is transmitted straight ahead into left field, while half the light arriving from third base is reflected into left field and half is transmitted straight ahead into right field. This mirror can be carefully placed so that the two beams headed for left field interfere destructively and the

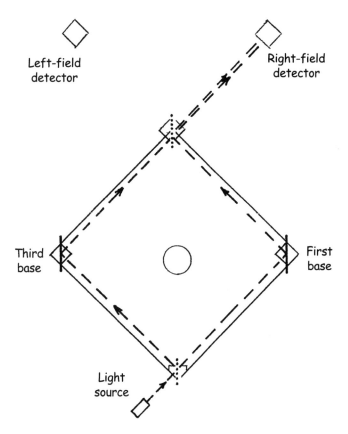

Figure 51. With a half-silvered mirror above second base, interference causes every photon to go to right field.

two beams headed for right field interfere constructively. Thinking classically, it's easy to predict what will happen. All the light will end up in right field (see Figure 51). The light waves, propagating along the base lines, will cancel each other along the path to left field and reinforce each other along the path to right field. What about single photons? As Wheeler correctly predicted, the detector in right field will sit there clicking repeatedly as photons arrive, while the detector in left field remains silent. There is only one conclusion: each single photon follows both paths at once. The fact that every photon ends up in right field can be explained only by means of a photon wave function (or amplitude) that divides and reconverges such that the photon wave can interfere constructively or destructively with itself.

So I have a choice. I can leave the space above second base clear, in which case my measurements show the path each photon followed. Or I can place the half-silvered mirror there, in which case I demonstrate that each photon followed both paths at once. Now enters *delayed* choice. I turn on my light source for only one nanosecond (a billionth of a second), in which time it emits, let's say, a few dozen photons. Then I relax for forty or fifty nanoseconds, scratching my head while I decide what to do next. Since a photon travels at a speed of about one foot per nanosecond, the photons from my little burst will be long gone from home plate by the time I make up my mind; but they will still be well short of second base. They will be en route.

Suppose I decide to find out what route each photon followed. That's easy. I leave the space above second base empty and count the arrival of photons in my left-field and right-field detectors. About half the photons will arrive at each detector. This suggests that by the time I made my decision, each photon was already "committed" to its path, either via first base or via third base. Suppose instead that I decide—*after* the photons are well on their way—to see if each photon followed both paths at once. I put the half-silvered mirror in place above second base. Miraculously, every photon then arrives at the right-field detector, indicating that each photon interfered with itself, requiring that each photon negotiated both paths at once.

I can do one more thing. I can leave the half-silvered mirror in place at second base and send a coach onto the field to block the path from home plate to first base (as in Figure 52). See if you can figure out what happens. Now no photon (or photon wave) can complete the path via first base. Every photon (or photon wave) that reaches second base must have proceeded via third base. This time the detectors click at equal rates. Each photon arriving at second base from third base has a 50-percent chance of being transmitted straight ahead to right field and a 50-percent chance of being reflected into left field. Interference has been eliminated. The photons again behave as single particles following specific paths.

With modern electronics, it is not hard to arrange for "decisions" to

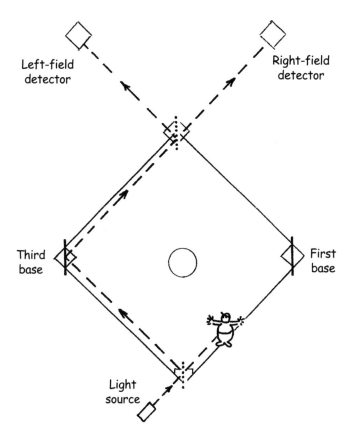

Figure 52. When the coach blocks the path to first base, the photons arrive in equal numbers in left field and right field.

be made in nanoseconds. If the experiment works in a laboratory room, there is no reason it shouldn't work on a baseball diamond. And as John Wheeler pointed out, there's no reason it shouldn't work over cosmic distances. Think of light from a distant quasar that can get to Earth by two possible paths. It can travel near Galaxy A and be deflected, let's say, to the left, putting it on track to reach Earth. (Deflection of light by gravity is now well established and often observed.) Or it can travel near Galaxy B and be deflected, let's say, to the right, putting it on a different track to reach Earth. If an astronomer points a telescope at Galaxy A, he or she sees those photons from the quasar which have passed near that galaxy. Looking toward Galaxy B, the astronomer sees photons which have passed near *that* galaxy. But the astronomer can do something else. He or she can put a half-silvered mirror at the place in the observatory where light arrives from both galaxies. Then (in principle) light that followed the two different paths will interfere, producing visible photons in only one direction (analogous to right field only). The astronomer's decision whether to look for a particular path or to look for interference between two paths can be made a billion years after the light left the quasar. To use Einstein's word, it is spooky. But it is real.

## Quantum Mechanics and Gravity

### Quantum Foam

I have already mentioned quantum foam, the roiling of spacetime at distances of $10^{-35}$ m and times of $10^{-45}$ s. In the world that scientists have studied so far, down to dimensions far smaller than the size of an atomic nucleus, gravity and quantum theory have nothing to do with each other. But as Wheeler pointed out, if you imagine probing far more deeply into the fabric of spacetime, you reach a point where the fluctuations and uncertainties that characterize particles take hold of spacetime itself. The smooth uniformity of the space around us and the ordered forward march of time give way to weird convolutions impossible to visualize. This is *Planck-scale physics* (from which Max Planck himself would no doubt have recoiled), now extensively explored by theoretical physicists.

### Strings

Down at the Planck dimension, or not far above it, are the hypothetical *strings* that another group of theorists study. According to string theory, what the physicist sees (and treats mathematically) as a particle—that is, an entity that exists at a mathematical point—is in fact a tiny oscillating string, either a line or a loop. The oscillations of the string give rise to the mass and other properties of the thing we call a particle. The mathematics of string theory is formidably difficult, and the theory has yet to account for the actual properties of the fundamental particles. Yet it is a most intriguing idea, with the potential to unite quantum theory and gravitational theory. One of its appealing features is that it replaces point particles—which are, both philosophically and mathematically, troublesome—with entities spread out over a small region of space. Imagine a charge located, literally, at a point. The electric field near this charge varies as the inverse square of the distance from it; so as you get closer and closer to the charge, the field grows without limit. At the charge, it is infinite. That's just one example of an infinity associated with point particles. Physicists will be happy if the things that we call particles and that seem to exist at points are eventually shown to occupy some region—albeit an incredibly small region—of space.

## Evaporating Black Holes

A *black hole* is often defined as an entity surrounded by such intense gravity that nothing, not even light, can escape from it. That was a good definition until 1974, when the celebrated British physicist Stephen Hawking, invoking quantum theory, showed that something *can* escape from a black hole, and indeed that a black hole can gradually evaporate, until all of its mass is radiated away. It all started with a remark that John Wheeler made to his graduate student Jacob Bekenstein. "Jacob," said Wheeler (I am paraphrasing), "if I set a cup of hot tea on a table and allow it to cool, I am committing a crime, for the transfer of heat from the hot tea to the cooler room increases the total disorder in the world, contributing to a general running down of the universe. But if I drop the hot tea into a black hole, I avoid punishment for the crime. That is an act that does not increase disorder." According to Wheeler, Bekenstein reappeared in his office a couple of months later, and said (again I paraphrase): "You can't escape punishment for your crime. A black hole possesses entropy (a measure of disorder), so when you toss the tea into a black hole, you are increasing the disorder in the universe and contributing to its demise as surely as if you'd left the tea on the table."

Hawking at first had difficulty accepting the idea published by Bekenstein—that a black hole possesses entropy. But when he did accept it, he began to think along the following lines. If a black hole has entropy, it must have temperature, and if it has temperature, it must radiate (just as the Sun radiates, and as every object, no matter how cold, radiates to some extent). But where is the radiation going to come from if nothing can escape from a black hole? Hawking realized that the incessant creation and annihilation of virtual particles in empty space could provide the answer. Normally, if a pair of particles—say, an electron and a positron—come into existence, they quickly annihilate, restoring momentary calm.* But if the pair is created right at the "horizon" of a black hole—that is, right at the boundary between the inner region from which nothing can escape and the outer region where escape is possible—one of the created particles can be sucked into the black hole while the other one flies away. The net result is that some small part of the black hole's energy is given to the escaping particle, and the mass of the black hole decreases slightly. Hawking's calculations showed that for a very massive black hole, the rate of "evaporation" is exceedingly slow, but that when a black hole has a small mass toward the end of its life—or had a small mass in the first place—the rate of evaporation can be swift. The black hole disappears in a final glorious burst. To date, no black hole has been observed ending its life in this way, but when it happens, punishment for Wheeler's "crime" will be meted out.

## Dark Matter

One of the current mysteries of the cosmos, which particle physics and quantum theory might help to solve, is the mystery of *dark matter*. For a long time, astronomers and cosmologists assumed that most of the matter in the universe is luminous. In our own solar system, there is some dark matter—namely, the planets and asteroids. But the total mass of all this dark matter is much less than the mass of the Sun. To a good approximation, the mass of the luminous Sun is the same as the mass of the entire solar system. An alien astronomer would not be far wrong if he or she (or it) took account of the mass of the Sun and ignored the mass of unknown and unseen objects in the Sun's vicinity. It was logical

---

* "Calm" is hardly the right word, since this creation-annihilation dance goes on everywhere at all times, making empty space quite a lively place.

to assume that other solar systems would be like our own—a massive central star surrounded by some puny chunks of dark matter. Luminosity, I should add, refers not just to visible light. Astronomers also "see" objects in the sky with infrared and ultraviolet waves, radio waves, and X rays. Dark matter is truly dark, emitting no detectable radiation of any kind. (Cold matter and massive black holes do radiate a bit, but not enough to be seen over cosmic distances.)

Even though it doesn't radiate appreciably, dark matter can show its presence by its gravitational effect. All matter, known and unknown, exerts and feels gravitational force. In recent years, astronomers have compiled convincing evidence that there is a great deal of dark matter in the universe, and they currently estimate its mass to be about six times the mass of luminous matter. One source of evidence is the rate of rotation of spiral galaxies. The stars in a galaxy don't rotate like the parts of a wheel. The rate of rotation of a particular star depends on its distance from the center of the galaxy and on the total mass inside its orbit. By studying the motion of stars in other galaxies, astronomers have concluded that the stars are being tugged by much more than just the other stars. Entire galaxies within clusters also move in a way that suggests there is much more mass than we can see "out there." This is a stunning conclusion. We thought we could "see" the universe, and now we find that we can't see most of it.

What is the dark matter? No one knows. This is one of the most compelling questions about our universe. The answer, when it is found, will shed light on other questions, such as: What, exactly, happened in the first moments after the Big Bang? The most prosaic idea is that dark matter consists of chunks of ordinary "stuff" (dust, rocks, planets, stars too small to glow). Another idea is that dark matter consists of neutrinos. The masses of neutrinos are not known, but they are probably massive enough to be acceptable candidates for dark matter. A definitely nonprosaic idea is that dark matter consists of new kinds of particles (called *exotic particles*, for lack of a better name) unknown in our laboratories. This would surely be the most exciting answer. Such hypothetical particles, if heavy (that is, at least as massive as a proton), must be weakly interacting, else they would have been seen by now. So they have acquired the name WIMPs, for "weakly interacting massive parti-cles." Will they eventually make an appearance in the tables of particles in Appendix B?

## Dark Energy

A popular supposition about our universe—a supposition supported now by a growing body of evidence—is that it will expand forever. Indeed, the evidence points to *accelerating* expansion. Let me explain why this is quite weird. Consider an object projected upward from the Earth's surface. If that object is a baseball or a toy rocket, it will rise to some height, come momentarily to rest, and fall back to the ground. If it is a spacecraft launched at a speed greater than what is called *escape ve-locity*, it will fly away forever, slowing until it reaches some final coasting speed out in interstellar space. If it is launched at a speed precisely equal to escape velocity, it will escape Earth's gravitational field, but barely. It will keep going forever, but will keep slowing until it approaches zero speed at a great distance from Earth. What all three of these possibilities have in common is that the projectile, once its propulsive force stops, *decelerates*. In no case will it *accelerate* away from Earth (once it is sent on its way).

Not so long ago, scientists assumed that the universe as a whole had the same three options. The material in the universe, launched by the Big Bang, could fly apart to some maximum extension, then turn around and collapse toward an eventual "Big Crunch." This behavior, analo-gous to the flight of the baseball or toy rocket, would occur, according to theory, if the total energy of the universe (mass energy plus any other forms of energy) was so great that gravity would stop the expansion and turn it around. Or, with a much smaller total energy and consequently weaker gravity, the universe could expand forever, slowing until it ap-proached a constant rate of expansion (a final coasting). At a certain *critical energy*, the universe would expand forever, but only barely, al-ways slowing but never stopping.

Then came the startling evidence, reported first in 1998, that the expansion is accelerating. Researchers examined the properties of a cer-tain kind of supernova in distant galaxies for which they could mea-sure both the speed and the distance. As expected for an expanding universe, those farther away were observed to be moving faster. But the data contained a surprise. The more distant the supernova—which means the farther the researchers peered into the past—the more its speed fell short of what would be expected for an expansion that is decelerating or is uniform. The rate of expansion now, it was implied, is greater than it was billions of years ago. In short, the expansion is accelerating. Theorists, as quick on the draw as the sheriff in a Western, had no trouble offering an explanation. They pinned the blame for accelerating expansion on *dark energy*.

What is dark energy? Well, first, it is *dark*, like dark matter—that is, invisible. It does not make its presence known through any kind of radiation that can be detected on Earth. Second, it is *energy*. Again, this seems to make it resemble dark matter, which is also a form of energy. But there are two big differences between dark matter and dark energy. First, dark matter is presumed to consist of bits of matter scattered through space—whether those "bits" be single particles or rocks or planets or even black holes. Dark energy, by contrast, is spread uniformly through space. It can even be thought of as a property of space. It is not congealed in bits. It is everywhere.

Second, dark matter is attractive and dark energy is repulsive. This needs a little explanation. Dark matter, like all other matter, is presumed to attract all matter gravitationally. Like ordinary matter, it tends to decelerate the expansion of the universe. Dark energy does not literally repel matter. Rather, it causes space itself to expand, and thus, indirectly, acts as if it is pushing matter apart. So dark and ordinary matter together *decelerate* the expanding universe, while dark energy *accelerates* it. According to theory, the *pull* of dark and ordinary matter weakens as time passes (as the parts of the universe get farther apart), whereas the *push* of dark energy remains constant. This means that acceleration, which already, after nearly fourteen billion years, seems to be triumphing over deceleration, will remain in the ascendancy forever.

One thing that ordinary matter, dark matter, and dark energy all have in common is their effect on the curvature of space. Whether space, in the large, has "positive" curvature, like the surface of a basketball, or "negative" curvature, like the surface of a saddle, or is "flat," like the Bonneville Salt Flats in Utah, depends on the total energy in the universe. The remarkable conclusion reached by cosmologists in the late 1990s (confirming the earlier hopes of many of them) is that the universe is flat. Moreover, evidence from a variety of sources suggests that the energy contributing to this flatness may be divided as follows: ordinary matter, 4 percent; dark matter, 23 percent; dark energy, 73 percent. Not only is our little Earth the tiniest possible speck in a vast universe of glowing matter—but even everything that glows is merely a small fraction of the total energy "out there."

Before leaving dark energy, I want to provide a bit of related history.

Einstein, following his original 1915 work on general relativity, became concerned because his equations predicted that the universe is *dynamic*—that is, either expanding or contracting. At the time, there was no evidence for a dynamic universe. Einstein, along with practically all other scientists, assumed that the universe was *static* (that its internal components rotated and rushed about, but sustained no overall growth or shrinkage). So he added to his equations a term that he called a *cosmological constant*. Its effect was to counter gravity—in effect, to act as antigravity—and thus to permit the existence of a static universe.

A decade later, the American astronomer Edwin Hubble discovered that the universe is in fact expanding. There seemed to be no need after all for the cosmological constant. Einstein's *original* equations, without the cosmological constant, could seemingly account beautifully for the whole aftermath of the Big Bang. Einstein reportedly called his introduction of that constant his biggest blunder.* And for nearly seventy-five years it disappeared from physics.

But accelerating expansion needs it. What Einstein called the cosmological constant (an all-too-bland term) is very likely part of what is now called dark energy. The new term conveys more clearly a physical idea of what is going on with this strange antigravitational flexing of space. If precisely tuned to just the right intensity, dark energy could, in principle, give rise to a static universe. It doesn't do that, however. According to present evidence, it is overpowering conventional grav-

---

* This remark by Einstein was reportedly made in a conversation with George Gamow (in German). Although Einstein never referred to his "biggest blunder" in print, it is one of his most quoted phrases.

ity and propelling the universe ever more rapidly outward. Einstein's "greatest blunder" was perhaps just another measure of his genius.

## An Eerie Theory

A rhyming word game that my wife and I used to play with our children goes by various names. We call it Stinky Pinky. One person gives a definition, such as "superior pullover," and the others try to guess the answer: "better sweater." Or they try to figure out that a "disillusioned mountaintop" is a "cynical pinnacle." And so on. What is "quantum mechanics"? It is an "eerie theory." In this book I have used fundamental particles as well as atoms and nuclei to illustrate this point. Physicists themselves often say that their heads swim when they think too hard about quantum mechanics. As I have stated earlier in this book, quantum mechanics is eerie not just because it violates common sense. It is strange for deeper reasons: it deals with unobservable quantities; it shows that nature's fundamental laws are probabilistic; it permits particles to be in two or more states of motion at the same time; it allows a particle to interfere with itself; it says that two widely separated particles can be entangled. All of this leads many physicists to believe that quantum mechanics, despite its long and unblemished record of success in accounting for subatomic phenomena, is incomplete. More and more physicists are agreeing with John Wheeler: "How come the quantum?" is a good question.

appendix a

# Measurements
# and Magnitudes

*Table A.1*  Large and small multipliers

| Factor | Name | Symbol | Factor | Name | Symbol |
|---|---|---|---|---|---|
| one hundred $10^2$ | hecto | h | one hundredth $10^{-2}$ | centi | c |
| one thousand $10^3$ | kilo | k | one thousandth $10^{-3}$ | milli | m |
| one million $10^6$ | mega | M | one millionth $10^{-6}$ | micro | $\mu$ |
| one billiion $10^9$ | giga | G | one billionth $10^{-9}$ | nano | n |
| one trillion $10^{12}$ | tera | T | one trillionth $10^{-12}$ | pico | p |
| one quadrillion $10^{15}$ | peta | P | one quadrillionth $10^{-15}$ | femto | f |
| one quintillion $10^{18}$ | exa | E | one quintillionth $10^{-18}$ | atto | a |

*Table* A.2    Table of measurements

| Physical quantity | Common unit in the large-scale world | Typical magnitudes in the subatomic world |
|---|---|---|
| Length | meter, m (a little more than a yard); also the centimeter, cm (0.4 inch), and the kilometer, km (0.6 mile) | Size of atom, about $10^{-10}$ m (0.1 nanometer, or 0.1 nm) Size of proton, about $10^{-15}$ m (1 femtometer, 1 fm) |
| Speed | meters per second, m/s (walking speed) or kilometers per second, km/s (speed of a bullet) | Speed of light, $3 \times 10^8$ m/s |
| Time | second, s (the swing of a pendulum); also hour, day, year | Time for a particle to cross a nucleus, about $10^{-23}$ s Typical lifetime of "long-lived" particle, about $10^{-10}$ s |
| Mass | kilogram, kg (mass of 1 liter of water) | Mass of two electrons, about 1 million eV, or 1 MeV Mass of proton, about 1 billion eV, or 1 GeV |
| Energy | joule, J (kinetic energy of a book that has fallen 10 cm) | Air molecule less than 1 eV, electron in TV tube about $10^3$ eV, proton in largest accelerator about $10^{12}$ eV (1 electron volt [eV] = $1.6 \times 10^{-19}$ J) |
| Electric charge | coulomb, C (lights a lamp for 1 s) | Magnitude of electron and proton charge = $1.6 \times 10^{-19}$ C |
| Spin | kg $\times$ m $\times$ m/s (person turning around) | Spin of photon = $\hbar \approx 10^{-34}$ kg $\times$ m $\times$ m/s |

appendix b
# The Particles

*Table B.1*  Leptons

| Name | Symbol | Charge (unit: $e$) | Mass (unit: MeV) | Spin (unit: $\hbar$) | Antiparticle | Typical decay | Mean life |
|---|---|---|---|---|---|---|---|
| *Flavor 1* | | | | | | | |
| Electron | e | $-1$ | 0.511 | 1/2 | $e^+$ | Stable | |
| Electron neutrino | $\nu_e$ | 0 | Less than $3 \times 10^{-6}$ | 1/2 | $\bar{\nu}_e$ | Stable (may oscillate to other neutrinos) | |
| *Flavor 2* | | | | | | | |
| Muon | $\mu$ | $-1$ | 105.7 | 1/2 | $\mu^+$ | $\mu \rightarrow e + \nu_\mu + \bar{\nu}_e$ | $2.2 \times 10^{-6}$ s |
| Muon neutrino | $\nu_\mu$ | 0 | less than 0.19 | 1/2 | $\bar{\nu}_\mu$ | Stable (may oscillate to other neutrinos) | |
| *Flavor 3* | | | | | | | |
| Tau | $\tau$ | $-1$ | 1,777 | 1/2 | $\tau^+$ | $\tau \rightarrow e + \nu_\tau + \bar{\nu}_e$ | $2.9 \times 10^{-13}$ s |
| Tau neutrino | $\nu_\tau$ | 0 | less than 18 | 1/2 | $\bar{\nu}_\tau$ | Stable (may oscillate to other neutrinos) | |

*Table B.2*  Quarks

| Name | Symbol | Charge (unit: $e$) | Mass (unit: MeV) | Spin (unit: $\hbar$) | Baryon number | Antiparticle |
|---|---|---|---|---|---|---|
| *Group 1* | | | | | | |
| Down | d | $-1/3$ | 5 to 8.5 | 1/2 | 1/3 | $\overline{d}$ |
| Up | u | 2/3 | 1.5 to 4.5 | 1/2 | 1/3 | $\overline{u}$ |
| *Group 2* | | | | | | |
| Strange | s | $-1/3$ | 80 to 155 | 1/2 | 1/3 | $\overline{s}$ |
| Charm | c | 2/3 | 1,000 to 1,400 | 1/2 | 1/3 | $\overline{c}$ |
| *Group 3* | | | | | | |
| Bottom | b | $-1/3$ | 4,000 to 4,500 | 1/2 | 1/3 | $\overline{b}$ |
| Top | t | 2/3 | 174,000 | 1/2 | 1/3 | $\overline{t}$ |

*Table B.3*  Some composite particles

| Name | Symbol | Charge (unit: $e$) | Mass (unit: MeV) | Quark composition | Spin (unit: $\hbar$) | Typical decay | Mean life |
|------|--------|--------|--------|--------|--------|--------|--------|
| *Baryons (which are fermions)* | | | | | | | |
| Proton | p | 1 | 938.3 | uud | 1/2 | none known | more than $10^{25}$ years |
| Neutron | n | 0 | 939.6 | ddu | 1/2 | $n \rightarrow p + e + \bar{\nu}_e$ | 886 s |
| Lambda | $\Lambda$ | 0 | 1,116 | uds | 1/2 | $\Lambda \rightarrow p + \pi^-$ | $2.6 \times 10^{-10}$ s |
| Sigma | $\Sigma$ | 1, 0, $-1$ | 1,189 ($+$ & $-$) 1,193 (o) | uus ($+$), dds ($-$) uds (o) | 1/2 | $\Sigma^+ \rightarrow n + \pi^+$ $\Sigma^o \rightarrow \Lambda + \gamma$ | $0.80 \times 10^{-10}$ s ($+$ & $-$) $7 \times 10^{-20}$ s (o) |
| Omega | $\Omega$ | $-1$ | 1,672 | sss | 3/2 | $\Omega \rightarrow \Lambda + \pi^-$ | $0.82 \times 10^{-10}$ s |
| *Mesons (which are bosons)* | | | | | | | |
| Pion | $\pi$ | 1, 0, $-1$ | 139.6 ($+$ & $-$) 135.0 (o) | $u\bar{d}$ ($+$), $d\bar{u}$ ($-$) $u\bar{u}$ & $d\bar{d}$ (o) | 0 | $\pi^+ \rightarrow \mu^+ + \nu_\mu$ $\pi^o \rightarrow 2\gamma$ | $2.6 \times 10^{-8}$ s ($+$ & $-$) $8 \times 10^{-17}$ s (o) |
| Eta | $\eta$ | 0 | 548 | $u\bar{u}$ & $d\bar{d}$ | 0 | $\eta \rightarrow \pi^+ + \pi^o + \pi^-$ | $5.6 \times 10^{-19}$ s |
| Kaon | K | 1, 0, $-1$ | 494 ($+$ & $-$) 498 (o) | $u\bar{s}$ ($+$), $\bar{u}s$ ($-$) $d\bar{s}$ & $\bar{d}s$ (o) | 0 | $K^- \rightarrow \mu^- + \bar{\nu}_\mu$ $K^o \rightarrow \pi^+ + \pi^-$ | $1.24 \times 10^{-8}$ s ($+$ & $-$) $0.89 \times 10^{-10}$ s (o) |

*Table B.4*  Force carriers

| Name | Symbol | Charge (unit: $e$) | Mass (unit: MeV) | Spin (unit: $\hbar$) | Antiparticle | Force that is carried |
|---|---|---|---|---|---|---|
| Graviton (hypothetical, never observed) | — | 0 | 0 | 2 | Self | Gravitational |
| W plus | W$^+$ | 1 | 80,400 | 1 | W$^-$ | Weak |
| W minus | W$^-$ | −1 | 80,400 | 1 | W$^+$ | Weak |
| Z particle | Z$^\circ$ | 0 | 91,190 | 1 | Self | Weak |
| Photon | $\gamma$ | 0 | 0 | 1 | Self | Electromagnetic |
| Gluon (set of 8 particles) | g | 0 (but 3 "color charges") | 0 | 1 | Self | Strong |

## appendix C

# Going for the Gold

Leptons, the subject of Chapter 3, have been responsible for quite a few Nobel Prizes. Here are the principal ones:

**Sir Joseph John Thomson** in 1906 for his "investigations on the conduction of electricity in gases" (his discovery of the electron).

**Prince Louis-Victor de Broglie** in 1929 for "his discovery of the wave nature of electrons."

**Carl David Anderson** in 1936 for his "discovery of the positron" (the antielectron).

**Cecil Frank Powell** in 1950 for "his discoveries regarding mesons" (showing the muon to be distinct from the pion).

**Sheldon Glashow, Abdus Salam,** and **Steven Weinberg** in 1979 for "the theory of the unified weak and electromagnetic interaction" (tying together the forces that govern neutrinos and those that govern charged particles).

**Leon M. Lederman, Melvin Schwartz,** and **Jack Steinberger** in 1988 for "the discovery of the muon neutrino."

**Frederick Reines** in 1995 for "the detection of the neutrino."

**Martin L. Perl** in 1995 for "the discovery of the tau lepton."

**Raymond Davis, Jr.** and **Masatoshi Koshiba** in 2002 for "the detection of cosmic neutrinos."

If Nobel Prizes for other work on the properties and behavior of electrons were included, the list would be longer. It would include **Niels Bohr** in 1922 for the quantum theory of electrons in the atom; **Arthur Compton** in 1927 for seeing and interpreting the scattering of photons by electrons; **Clinton Davisson** and **George Thomson** (son of J. J.) in 1937 for using crystals to diffract electron waves; **Wolfgang Pauli** in 1945 for discovering that no two electrons can be in the same state of motion at the same time; **Willis Lamb** in 1955 for precision measurements of electron energies in the hydrogen atom; **Polykarp Kusch** in 1955 for precisely determining the magnetic properties of the electron; **Robert Hofstadter** in 1961 for using electrons to probe the interior of protons and neutrons; **John Bardeen, Leon Cooper,** and **J. Robert Schrieffer** in 1972 for the theory of superconductivity (the frictionless motion of electrons in some materials); and **Jerome Friedman, Henry Kendall,** and **Richard Taylor** in 1990 for using electron beams to reveal quarks in nucleons.

# Index

*Page numbers followed by the letter n refer to footnotes. Numerals in italics refer to pages with illustrations.*

Absolute zero, 109
Acceleration of cosmic expansion, 244–245
Accelerator(s)
  to achieve small wavelengths, 211
  to create new particles, 212
  creation of mass in, 21
  energies of, 21
  linear, 212
  wavelength of particles in, 12
Alley, Carroll, 235
Alpha decay, 37–39, 123, 203
Alpha particle. *See* Helium nucleus
Alpha rays, 32, 33
Alsos, 137n
Anderson, Carl, 35, 257
  and discovery of positron, 34–35, 36
Angular momentum
  conservation of, 154, 161–163
  quantization of, 25
  quantized orientation of, 26
  spin and orbital distinguished, 24, 25
  *See also* Orbital angular momentum; Spin
Annihilation. *See* Creation and annihilation
Antielectron. *See* Positron
Antigalaxies, 226
Antihydrogen, 224–225

Antimatter
  as an energy "source," 223–225
  first predicted, 33
  storage of, 224
Antiparticles, 35
  instability of, 225
Anti-periodic table, 226
Antiprotons
  in cosmic radiation, 226
  stability of, 225
Apartment-house model, 142–143, 145, 148–149
Aristotle, 154
Atomic number, 5
Atoms
  size of, 1n, 198–201
  speed of, 14

Background radiation, 16
Bahcall, John, 62n
Ball bearing *vs.* electron, 100
Bardeen, John, 258
Baryonic charge. *See* Baryon number
Baryon number, 46
  conservation of, 69, 164
Baryons, 71
  decay of, 72–73
  defined, 46
  origin of word, 46

Beat note, 61–62
Becquerel, Henri, 32, 36, 38
Before-and-after views, 155
Bekenstein, Jacob, 241
Beta decay, 39–45
Beta rays, 32, 33
Big Bang, 177, 226, 243, 246
Big Crunch, 244
"Biggest blunder," 246, 247
Binnig, Gerd, 124, 125
Bit, classical, 232
Black-body radiation, 98, 144
    See also Cavity radiation
Black hole(s)
    coinage, 13n, 98, 233
    and conservation laws, 147
    entropy of, 241
    evaporation of, 241–242
    temperature of, 241
Bohr, Niels, 106, 118n, 199, 258
    and angular-momentum
        quantization, 101
    and challenge of quantum theory, 98,
        220
    1913 paper, 96, 106–108, 199
Bohr atom, 106–108, 134
Bohr-Einstein debates, 196, 220–221
Born, Max, 112, 113, 118, 185n, 204,
    205
Bose, Satyendra Nath, 143–146,
    185
Bose-Einstein condensate, 145, 147–
    150, 222, 230
    early data for, 149
    practical potential of?, 149–150
Bose-Einstein statistics, 145
Bosons, 47, 143–145
    composite, 132
    wave functions of, 152
Branching ratio, 117, 118, 205
Brown, Julian, 233n
Brush, Stephen, 193n

$c^2$ as energy per unit mass, 19
Calorie, 21
Carbon atom, 139
Cathode rays, 30
Cathode-ray tubes. See CRTs
Cavity radiation, 92–94, 94, 95, 97,
    144
CERN, 75, 212n, 224–225
Chaos, 114n
Charge, 22–24
    conservation of, 163
    fractional, 23
    sign convention for, 22
Charge conjugation symmetry (C), 166
    partial conservation of, 175–178
Charge conservation, 23
Charge of electron, 10, 101
Charge quantization, 23, 100–101
Classical physics, 3–4, 105–106
Clusters of galaxies, 243
Cobalt 60, 172–174, 173
Coin flipping, 113–114, 115–116
Colliding beams, 52
Collision
    proton-neutron, 169
    proton-proton, 146–147, 156, 169
Color, 70–71, 79–80
    conservation of, 70, 165
    quantization of, 103–104
Color charge. See Color
Colorlessness, 104
Common sense, limits of, 4–5
Complexity
    in large-scale world, 59
    layers of, 59
Complex number, 151n, 204n
Composite particles, 71–73
    properties of, 255
Compton, Arthur, 144, 184
Compton effect, 184, 185n
Condon, Edward, 37, 39
Conduction electron, 110

Confinement and energy quantization, 109–110
Conservation laws, 153
  absolute, 159–165
  and allowed behavior, 157
  centrality of, 154–155
  links to invariance principles, 159
  links to symmetry principles, 180, 181, 183
  and obligatory behavior, 157
  partial, 168–178
  short history of, 154
  sufficiency of?, 183
  *See also entries for specific conserved quantities*
Constant of proportionality, 19
Cooper, Leon, 258
Cornell, Eric, 147, 148
Corpuscle. *See* Photon(s)
Correspondence principle, 106–107
Cosmic background radiation, 76
Cosmic radiation, 45
  and discovery of positron, 34
  at high energy, 21
  primary, 2, 45n, 226
Cosmological constant, 246
Coulomb, Charles, 23
Coulomb (unit), 23, 251
Cowan, Clyde, 43
CP symmetry, 176–178, 226
Creation and annihilation, 65
  central to quantum physics, 43
  at every vertex, 86–87
  of hydrogen atoms, 110
Critical energy of universe, 244
Cronin, James, 176, 226
Cronkite, Walter, 5
CRTs, 30–31, 31n, 32
Curie, Marie, 32, 35
Curie, Pierre, 32
Cyclotrons, early, 21

Dark energy, 244–247
  nature of, 245
Dark matter, 242–244
  mass of, 243
Davis, Raymond, Jr., 257
Davisson, Clinton, 185, 186, 187, 193, 258
de Broglie, Louis-Victor, 184–185, 206, 257
  and matter waves, 199–200
de Broglie equation, 189–192, 227
Decay, 37
  as explosive event, 116
Degree of freedom, 136
Delayed choice, 233–240
  and cosmic distances, 239–240
Deuteron, 63–64
  mass of, 104
Deutsch, David, 233
Diffraction, 193
  of electron waves, 187
  of light, 194
  of single photon, 198
  of water waves, 12, 211
Dirac, Paul, 34, 34, 35, 58
  and Fermi-Dirac statistics, 138
  his coinage of boson, 145
  his coinage of fermion, 145n
  and prediction of positron, 33
  as thinker, not talker, 33n
Dirac equation, 33
Direction invariance, 158
Dishonesty in science, 142n
Double-slit experiment, 194–198, 195, 197
  with electrons, 195–196
Downhill rule, 125–127

$E = hf$. *See* Planck-Einstein formula
$E = mc^2$, 17, 19
  *See also* Mass-energy equivalence
Egyptians, 27

Eigenvalue, 203
Einstein, Albert, 20
  and Bose-Einstein statistics, 145–146
  and Bose's paper, 143–145
  and cosmological constant, 246
  and invariance of all laws, 159
  and mass-energy equivalence, 19
  quotation of, *129, 130*
  reaction of to Feynman's idea, 219
  and the speed of light, 28
  *See also* General relativity; Special
    relativity
Electric charge. *See* Charge
Electric force in large-scale world, 74
Electromagnetic interaction, 76–79
Electromagnetism, 92
Electron, 2, 30–35
  diffraction and interference of, *187*
  discovery of, 31
  hypothetically spinless, 138–139
  limit on lifetime, 163
  mass-to-charge ratio, 31
  as practical workhorse, 23
  properties of, 253
  simplicity of, 99
  speed of in atom, 189
  stability of, 163
  track of, 49
  wave nature of, 185
Electron neutrino, 35–45
  properties of, 253
Electron volt (unit), 10
Electroweak interaction, 76–78
Element(s), 5
  heavy, 142
Emulsion detector, 48, *49,* 56
Energy, 18–22
  conservation of, 18, 160, 183
  as contributor to mass, 68, 104
  importance of concept, 18
  levels in H atom, 210
  levels of particle in a box, 208, 209

many units for, 21
  quantization of, 105–110
  storage of, 223
  *See also entries for specific kinds of en-
    ergy*
Energy change and photon frequency,
  96
Entanglement, 231–233
Entropy of black hole, 241
Escape velocity, 244
Eta, 72
  properties of, 255
Ether, 188
Euripedes, 153
Evaporation of black holes, 241–242
Event, 84–85
Exclusion principle, 100, 133, 150
Exotic particles, 243
Expansion of universe
  accelerating, 244–245
Exponential decay, 120–121, *121,* 128
Exponential notation. *See* Scientific
  notation

Faith in simplicity, 57–59
Femtometer, 9
Fermi, Enrico, 9n, 42, *42*
  and beta-decay theory, 75
  and Fermi-Dirac statistics, 138
Fermi (length unit), 9n
Fermions, 46–47, 133–143
  composite, 132
  spin of, 101
  wave functions of, 152
Feynman, Richard, 47, 81, *82,* 98–99
  and backward-in-time paths, 85, 88
  *QED,* 99n, 219n
  and sum over histories, 218–219
Feynman diagrams, 81–91
Field, 188
Fission by slow neutrons, 213
Fitch, Val, 176–177, 226

Flatness of universe, 246
Flavor(s)
  conservation of, 48, 64
  of quarks, 169–170
  why three?, 64–66
Force, electric, 22
  in nucleus, 22
  strength relative to gravity, 22
Force, gravitational, 17
Force carriers, 73–81
  defined, 46
  properties of, 256
Ford, Adam, 131n
Franklin, Benjamin, 22, 154
Free-fall equation, 156
Fresnel, Augustin, 193, 194
Friedman, Jerome, 258
Fusion energy in Sun, 140
Fusion furnace, 123

Galaxies, spiral, 243
Galileo Galilei, 154, 159
Gamma decay, 39
Gamma rays, 32, 33n, 39
Gamow, George, 18n, 37, 39, 246n
Garwin, Richard, 174
Geiger, Hans, 118
Geiger counter, 118–119
Gell-Mann, Murray, 6, 7
General relativity, 58, 59
Geon, 132n, 233
Germer, Lester, 185, 186, 187, 194
Gilbert and Sullivan, 131
Glashow, Sheldon, 76, 78–79, 79, 257
Gluons, 72, 79–81
  exchange of, 90
  interacting with each other, 80
  lifetime of, 15
  nature of force, 80
  not seen alone, 81
  properties of, 256

Gödel, Kurt, 24
Golden age of physics, 3
Goudsmit, Samuel, 25, 136, 137, 137n
Granularity, 99
  related to waves, 205–210
Gravitational force, 17
  decelerating effect of, 244
Gravitational interaction, 73–74
Graviton, 6n, 73
Ground state, 106, 108, 109, 116
Gurney, Ronald, 37, 39

Hadrons, 46, 71
Half-life, 37, 120–121, 121
  compared with mean life, 121n
  measurement of, 122
Half-silvered mirror, 235, 236, 237
Handedness. See Neutrinos
Harmonics, 206, 207
Hawking, Stephen, 98n
  and evaporation of black holes, 241
h-bar, 26, 27n
Heavy hydrogen, 63
Heavy water, 62, 63
Heisenberg, Werner, 203, 213
  See also Uncertainty principle
Helium nucleus, 140, 141
Heraclitus, 153
Higgs, Peter, 168
Higgs particle, 6n, 168
Hilbert, David, 181n
Hiroshima, 192, 224
Hofstadter, Robert, 211–212, 258
Holton, Gerald, 193n
Homogeneity of space, 157–158, 180, 181–182
Horizon of black hole, 242
"How come the quantum?", 98, 247
Hubble, Edwin, 246
Huygens, Christiaan, 154
Hydrogen as energy "source," 223–224

Hydrogen atom
    lowest-energy state of, 204–205
    quantum jump in, 126–127

Identity of particles, 100, 152
Improbable events, 225
Indistinguishability, 151
Inertia, 16
Inertial frame of reference, 158–159
Inflation, cosmic, 226n
Initial conditions, 114, 156
Interference, 193
    in double-slit experiment, 195
    of electron waves, *187*
    of light, *194*
    of matter wave, *191*
    of photon with itself, 198, 237–238
Interference fringes, 195
International Centre for Theoretical
    Physics, 78
Invariance principles, 157–159, 180
    absolute, 159–165
    links to conservation laws, 159
    partial, 168–178
    *See also entries for specific invariant*
        *quantities*
Inverse-square law, 58, 59
*Iolanthe,* 131
Island of stability, 141–142
Isospin, 170–171
Isotopes, 6
Isotropy of space, 158, 182
"It from Bit," 233–235

Jensen, Hans, 141
Joliot-Curie, Frèdèric, 34–35
Joliot-Curie, Iréne, 34–35
Jordan, Pascual, 204
Joule (unit), 21, 251

Kamiokande detector, 60, 61
Kaons, 72, 171n

and CP violation, 176
    properties of, 255
Kendall, Henry, 258
Kepler, Johannes, 154
Kepler's second law, 154, 155
Ketterle, Wolfgang, 147n, 148
Kilogram, 27
Kilowatt-hour, 21
Kinetic energy, 17, 18–19
    classical formula for, 19n
    of electron in atom, 200
Koshiba, Masatoshi, 257
Kramers, Hendrik, 118n
Kusch, Polykarp, 258

Ladder of energy states, *117*
Lamb, Willis, 258
Lambda
    decay of, 170
    properties of, 255
Large Hadron Collider, 212n
Laser cooling, 147
Law of compulsion, 155–156
Law of prohibition, 156–157
Lawrence, Ernest, 142n
Lead, 208, 141
Lederman, Leon, 49, 50, 51, 174, 257
Lee, Tsung-Dao, 171, 173–174
Length, 10–13
    smallest probed, 11
Lepton number, 47
    conservation of, 164–165
Lepton(s), 46
    families, 29
    flavors, 29
    origin of word, 29
    as point particles, 30
    table of properties, 253
    and weak interactions, 30
Lewis, Gilbert, 144n
Lifetime. *See* Half-life; Mean lifetime
Light speed. *See* Speed of light

Light-year, 9
Line spectrum, 105
Liquid droplet model, 140
Lithium atom, 139
Location invariance, 157–158
Logic gate, 232
Longitudinal wave, 187n
Lorentz invariance, 158–159
Luminous matter, 242–243

Mach number, 9
Magic numbers, 141–142
Magnetic trapping, 147
Mass, 16–18
    conservation of, 154
    lack of regularity, 105
    measurement of, 17
    quantization of, 104–105
    uncertainty of, 66
Mass energy, 17, 18
Mass-energy equivalence, 17, 191, 192
Matter-antimatter asymmetry, 177–178
Mayer, Maria, 141
Mean lifetime, 115
    compared with half-life, 121n
Measurement, quantum, 232
Measurements, table of, 251
Mesons, 46, 71
    decay of, 72
    origin of word, 46
    quark composition of, 71
Mesotron, 46
Metals, 143
Mirror reversal symmetry. See Parity
    symmetry (P)
Mixing of states. See Superposition
Molecules, speed of, 14
Momentum, 21n
    classical definition of, 189
    conservation of, 160–161, 182
    for massless particle, 189n
    of a particle in a box, 209

Multipliers, table of, 250
Muon neutrino, 48–51
    distinctness of, 51
    properties of, 253
Muon(s), 45–48
    in cosmic radiation, 16, 120n
    decay of, 88, 89, 116, 146, 164
    hypothetical decay of, 47
    mean life of, 47
    properties of, 253
    track of, 49

"Natural" units, 27–28
Neutrino oscillation, 61–62, 64, 165
Neutrinos
    handedness of, 166, 167, 174, 175
    mass of, 56–57, 59–64
    from pion decay, 49
    as possible dark matter, 243
    solar, 65
    speed of, 14
    weakness of interaction, 44
    See also Electron neutrino; Muon
        neutrino; Tau neutrino
Neutron(s), 5
    decay of, 88, 89, 146, 162
    discovery of, 42
    mean life of, 16, 70, 72
    properties of, 255
    quark composition of, 71
    stabilization of, 70
    thermal, 213
Newton, Isaac, 154, 181, 193n
    and momentum, 21n
Newton's third law, 181–182
Nobel Prizes for work on leptons, 257–
    258
Noether, Emmy, 180–181
Nonlocalizability related to waves,
    211–213
Notation, exponential. See Scientific
    notation

Notation, scientific. *See* Scientific notation

Nuclear reactor. *See* Reactor, nuclear

Nuclear waste, 122–123

Nuclei, atomic, 140–143
    shell structure of, 140–142
    size of, 10

Nucleons, 46, 170

Omega properties, 255

One-particle decay prohibited, 160–161

Orbital angular momentum, 162–163
    orientation of, 102
    quantization of, 101

Oscillation, neutrino, 61–62, 64, 165

Oscilloscopes, 31n

Overtones, 206, 207

Oxygen atom, 139

Parity symmetry (P), 166
    partial conservation of, 171–176

Parsec, 9n

Particle in a box, 207–209, 208, 209

Particles
    collisions of, 21
    plethora of, 5
    relations among, 111
    *See also entries for specific particles*

Pauli, Wolfgang, 40, 41, 56, 258
    and exclusion principle, 134, 136
    and fourth quantum number, 136
    suggestion of "neutron," 40–42, 133
    and theory of fermions, 138
    *See also* Exclusion principle

PC symmetry. *See* CP symmetry

Pentaquark, 71n

Periodic table, 5, 136
    and the exclusion principle, 140
    and quantum numbers, 139

Perl, Martin, 51–52, 52, 54, 257

Permissiveness, 156–157

Perpetual motion, 222–223

Photoelectric effect, 185, 192–193

Photon(s)
    early history of, 144
    early view of, 3
    exchange of, 86
    how numerous, 76
    properties of, 256
    self-interference of, 237–238
    speed of, 14

Pion(s), 45
    charged, decay of, 116, 117, 147, 166, *175, 177*
    composition of, 72
    lifetime of, 116
    mass of, 72
    neutral, decay of, 126, 161, 162, 193
    properties of, 255
    track of, 49

Planck, Max, 13n, 92, 93

Planck-Einstein formula, 39, 95, 217

Planck formula, 94, 144

Planck length, 13
    coinage, 13n

Planck scale, 13, 240
    coinage, 233

Planck's constant, 95, 99
    as fundamental constant, 28, 191
    as natural unit, 27
    setting scale of subatomic world, 191
    and size of atoms, 201
    small size of, 101
    ubiquitous in quantum theory, 213
    as wave-particle link, 189

Plum-pudding model, 198–199

Plutonium 239, 140

Positron
    annihilation of, 33, 146, 162
    as backward-in-time electron, 87–88
    discovery of, 34
    stability of, 225
    track of, 36

Positronium, 67n

Potential energy of electron in atom, 200
Powell, Cecil, 48, 257
Power, 21n
Primary cosmic radiation, 2
Probability
  classical vs. quantum, 118
  exactness in theory, 113–114
  fundamental, 114
  of ignorance, 114, 128
  linked to waves, 201–205
  of location, 205
  measurement of, 115, 202–203
  in quantum mechanics, 112
  in radioactivity, 37
  role in particle detection, 44–45
  of time, 205
Proton, 2
  limit on lifetime, 164
  mass of, 18
  possible instability of, 70n
  properties of, 255
  quark composition of, 71
  size of, 11, 13
  stability of, 69–70
  structure of, 212
Proton-neutron collision, 169
Proton-proton collision, 146–147, 156, 169

Quanta, coinage, 94
Quantization
  classical, 185, 206
  of orientation, 102–103
Quantum computing, 220, 232–233
Quantum foam, 13, 59, 240
  coinage, 13n, 233
Quantum mechanics, 2
  and gravity, 240–244
Quantum numbers, 134, 135–136
  and the periodic table, 139

Quantum permissiveness, 156–157
Quantum physics
  harnessing, 220
  its role in shaping the large-scale world, 222
  why it is troubling, 221
Quarks, 46, 67–71
  baryon number of, 69, 70
  combinations of, 67, 70
  "discovery" of, 6
  exchanging gluons, 90
  flavors of, 169–170
  fractional charges of, 69
  names of, 69
  not seen alone, 81
  properties of, 254
  strong interaction of, 67
  uncertain masses of, 67, 68
Qubit, 232

Rabi, I. I., 47
Radiation
  classical, by atom, 199
  See also Background radiation; Black-body radiation; Cavity radiation; Cosmic background radiation; Cosmic radiation
Radioactivity, 36–40
  as catastrophic change, 119
  discovery of, 32
  energy released in, 36–37
Randomness, 119, 120
Reactor, nuclear, 42, 43
Reines, Frederick, 43, 44, 257
Relativity. See General relativity; Special relativity
Rigidity (of tracks), 17
Rohrer, Heinrich, 124, 125
Rotational symmetry, 179
Rubidium 85 and 87, 148
Russell, Bertrand, 183
  The ABC of Relativity, 183n

Rutherford, Ernest, 107–108, *108*
  and alpha particle scattering, 11
  and exponential decay, 120
  and nuclear atom, 105, 199
  and radioactivity, 118
  and studies of radioactivity, 32

Salam, Abdus, 76, *77*, 77–78, 257
Scanning tunneling microscope. *See*
  STM
Scattering
  of alpha particles, 11
  electron-electron, 85–86, *86*
  of electrons by protons, 11
  photon-electron, 144, 184, 202
  *See also* Collision
Schrieffer, J. Robert, 258
Schrödinger, Erwin, 185, 203, 204
Schrödinger equation, 203, 204, 206
Schwartz, Melvin, 49, 50, 51, 257
Science fiction, 224
Scientific notation, 8–9
  multiplication and division in, 9
Seaborg, Glenn, 142n
Second, 27
Self-reinforcement of waves, 208
Shells, 135–136
Shock waves, 188n
Sigma properties, 255
Sine wave, 214–215
SLAC, 51–52, 53
  accelerator energy, 212
Slater, John, 118n
SNO, 61–64, *63*
Soddy, Frederick, 118
Sodium 23, 148
Space
  homogeneity of, 157–158, 180, 181–182
  isotropy of, 158, 182
Space contraction, 159
Space curvature, 245

Space-inversion symmetry, 179
  *See also* Parity symmetry (P)
Space map, 82–83, *83*
Spacetime map, 83–84, *84*
Special relativity, 2, 158, 159
Speed, 13–15
Speed limit, 13
Speed of light, 13
  as fundamental constant, 28
  human awareness of, 14
  as natural unit, 27
  as speed limit, 13
Spent fuel, 123
Spin, 24–27, 162–163
  of electron, 137–138
  measurement of, 26
  nuclear, 40
  of particles, 24–25
  quantization of, 100–103
  quantized orientation of, 26, *103*
Standard model, 6
Stanford Linear Accelerator. *See* SLAC
Starship Enterprise, 14–15, 224
*Star Trek,* 225
State of motion, 134–135
  classical, 16
Stationary state, 106, 108, 116
Steinberger, Jack, 49, 50, 51, 257
STM, 124, *125*
  image created by, *126*
storage ring, 52, 53
String theory, 64–65, 240–241
Strings, 13
  finite size of, 241
Strong interaction, 79–81
  most constrained, 168
Subatomic world, 1, 4, 8
Sub-subatomic world, 28, 220
Sudbury Neutrino Observatory. *See*
  SNO
Sumerians, 27
Super Kamiokande, 61

Superconductivity, 222
Superfluidity, 222
Superposition
  of amplitudes, 219, 228
  as intermingling, 229
  of many waves, 216
  of momenta, 227
  and probability, 228, 230
  of separated entities, 231
  of sine waves, 215–216
  of spin directions, 228–229, 230
  of states, 227–228
  and the uncertainty principle, 213–217
Symmetry, 178–183
  discrete and continuous, 181
  links to conservation laws, 180, 181, 183
  *See also entries for specific symmetries*

Tachyon, 15
Tau, 51–56, 54
  large mass of, 30
  properties of, 253
Tau neutrino, 56
  properties of, 253
Taylor, Richard, 258
TCP symmetry, 165–167
Temperature of black hole, 241
Tevatron, 212
Thermodynamics, 92
Thomson, George, 185, 194, 258
Thomson, J. J., 30–32, 31, 198, 257
Time, 15–16
Time dilation, 159
Timekeeping, 216–217
Time reversal symmetry (T), 166
  violation of, 176–178
Time-reversed path, 85
Time symmetry, 183
Translational symmetry, 178, 178–179
Transmutation, 119, 120

Transverse wave, 187n
Tunnel diode, 124–125
Tunneling, 37–39, 123–124, 203

Uhlenbeck, George, 25, 136, 137, 137n
Uncertainty principle, 39, 66
  classical version of, 217
  effect on timekeeping, 216–217
  momentum-position form of, 213–214
  statement of, 213
  and superposition, 213–217
  time-energy form of, 122, 216
  as tool of measurement, 122, 216
  and wave nature of matter, 214
Units, 9, 27–28
Universe
  accelerating expansion of, 244–245
  flatness of, 246
  lifetime of, 16
  mass of, 18
  size of, 12–13
Unobservable quantities, 150–151, 152
Uphill change, 127, 160
Uranium 235, 140
Uranium atom size, 10n
Uranium nucleus, 6

Vertices, three-prong, 86, 91
Violin string, 206, 207
Virtual particles, 58, 59, 65, 242

W particle, 75, 212
  properties of, 256
Warp drive, 14–15
Water wave, 186–187
Wave amplitude. *See* Wave function
Wave(s)
  banishing of, 218–219
  and granularity, 205–210
  localized, 188
  nature of, 200

Wave(s) (*continued*)
    and nonlocalizability, 211–213
    and probability, 201–205
    self-reinforcement of, 190, 191, 208
    speed of, 88
    as tools, 211
Wave function, 150–152, 204
    of electron in H atom, 205, 206
    symmetric and antisymmetric, 152
Wavelength
    of electron, 190
    of people, 190
Wave-particle duality, 184, 186, 196,
    217–218
Weak interaction, 75
    experienced by neutrinos, 43
    universality of, 77
Weightlessness, 17
Weinberg, Steven, 76, 78, 78, 257
Weinrich, Marcel, 174
Wheeler, John A., 59, 198, 221, 233,
    234, 240
    autobiography, 218n, 235

and backward-in time paths, 85, 88
coinages, 13n
and entropy of universe, 241, 242
naming black hole, 98
and sum over histories, 218–219
Wieman, Carl, 147, 148
Wigner, Eugene, 141
WIMPs, 244
World line, 84–85
Wu, Chien-Shiung, 171, 173

Yan, Helen, 97
Yang, Chen Ning, 171, 174
Young, Thomas, 193, 194
Yukawa, Hideki, 45
    and exchange particle, 76

$Z^\circ$ particle, 65–66, 212
    properties of, 256
Zero-point energy, 108–109
Zweig, George, 6

# THE QUANTUM WORLD